Part I
Design and Innovation

The Open University

Block 2

Engineering by design

This publication forms part of an Open University course T173 *Engineering the future*. Details of this and other Open University courses can be obtained from the Student Registration and Enquiry Service, The Open University, PO Box 197, Milton Keynes MK7 6BJ, United Kingdom: tel. +44 (0)845 300 60 90, email general-enquiries@open.ac.uk

Alternatively, you may visit the Open University website at http://www.open.ac.uk where you can learn more about the wide range of courses and packs offered at all levels by The Open University.

To purchase a selection of Open University course materials visit http://www.ouw.co.uk, or contact Open University Worldwide, Michael Young Building, Walton Hall, Milton Keynes MK7 6AA, United Kingdom for a brochure. tel. +44 (0)1908 858793; fax +44 (0)1908 858787; email ouw-customer-services@open.ac.uk

The Open University
Walton Hall, Milton Keynes
MK7 6AA

First published 2001. Second edition 2002. Third edition 2007.

Edited and designed by The Open University.

Typeset by SR Nova Pvt. Ltd, Bangalore, India

Printed in the United Kingdom by Thanet Press Ltd, Margate

ISBN 978 0 7492 2346 5

3.2

Contents

1 Design and designing

1.1 Aims

- To present and illustrate design practice.
- To disentangle the complex relationships between products and processes.
- To identify innovation in a wide variety of designed objects.
- To illustrate the influence of principles, context, method, and practices on the products of designing.
- To show models of the design process.
- To present examples of engineering design.

1.2 Introduction

In Block 1, you saw that design was important to all engineering activity. In this block, we shall explore the role of design in more detail.

Design is everywhere. Look around you, and consider the objects you see. For example, in my office I can see my computer, a telephone, a pen, a coffee mug, my sunglasses, a stapler, my wallet, my diary, a one pound coin, a postcard, carpet tiles, a desk, a chair, window blinds, a jacket, a radiator, a strip light, and much more besides. All these objects are the result of a number of decisions which have been made by someone – either an individual or, more likely, a team of people. The designing of the material world is thus a complex and multifaceted activity involving a variety of human capabilities. It is this complexity which is explored in this section. Straight away we can see that we need to make a distinction between the human capability for *designing* and the output of that process: the *designs* which surround us in their many forms. In this field, authors often refer to both the process and the product as 'design'. Watch for this and try to work out which they are referring to.

Some of the products that are considered in the block may be unfamiliar to you, and you may not know the detailed principles of their operation. You should still be prepared to think broadly about the challenges that faced the designer, though. Do not be put off by an unfamiliar product. You can be sure that the assignment material related to this block will not assume an advanced knowledge of these products and their manufacture.

To help give you an appreciation of the wide spectrum of contexts in which designing is undertaken and the variety of designs which emerge, our case studies will be taken from a variety of fields. The list below gives some of the examples we shall be looking at:

designing a folding bicycle

designing a human-powered aircraft

designing a rescue stretcher

designing a kettle

designing yachts and their hulls.

These examples will be used to illustrate the process of design, and the effect that early design decisions can have on the final product. Technicalities will only be introduced to illustrate the design story. You will not be expected to be a competent designer yourself after studying this block!

In addition to the wide variety of contexts in which designing occurs there is also great variation in the types of knowledge required by designers. Design teams are rarely static in their composition, and will rarely rely on the skills of designers alone. Specialist contributions will be sought throughout the whole process of designing: for example, advice on a potential market, information on a new material or results from the testing of a prototype. However, at the core of a design team will be people who are able to interpret information. They will have developed a certain blend of skills and knowledge which they use to combine and transform information into creative, new products. Interestingly, they might describe themselves as an engineer, innovator, designer, architect, inventor etc., but their titles are really of no importance here. I am more concerned with exploring those human capabilities which make people good at designing and innovating. You may wish to develop your own capability at this, but it can take many years of practice.

One thing all those objects in my office have in common is that they were all made in large numbers. This is not to say that you cannot be designing if you are not planning and specifying for mass manufacture. Indeed, one of the important messages in this course is that we can all find ourselves designing to some extent during our daily lives. Many people who have acquired a powerful ability for designing use it to make one-off designs – for example craft workers in wood or silver, or designers for the theatre. However, this block is biased towards the particular demands of designing for mass manufacture and mass consumption.

The term *design* can be, and indeed is, used to describe the creative output of various professions such as jewellers, architects, boat builders, and those people devising new television adverts. So a study of design would not be complete without a study of *designing*, and this part of Block 2 will guide you through both. It will look at the designs of our manufacturing culture including bridges and architecture as well as consumer products. It will also examine the process of designing, including a critical appraisal of some of the accepted models of design. The term *innovation* is widely used today and this block reflects on what constitutes innovative design and innovative designing. Design is an essential part of engineering, and, in a competitive world, innovation is an essential part of design.

1.3 Problems and solutions

One way to look at design is to consider it as a problem-solving activity. For example, a person designing the interior of a house has to solve many problems such as how to make it functional in an appropriate way (you don't, presumably, want your bed next to the cooker), how to make it attractive, how to make it comfortable, and how to achieve all this on a given budget. The designer needs to ask: 'Whom am I designing for?' An interior for one client may be very different from one designed for another client. Also, an interior which is intended to enhance the saleability of a property may be very different from one aimed at the owner's personal preferences. The design team responsible for a new motor car also has complex problems to solve including achieving a broad market appeal as well as meeting the required performance. There are various modelling techniques which designers use to help them understand problems and generate solutions.

Generally design problems comprise several factors. Those factors that are concerned with how people use, understand or interact with designs could be called 'human factors'. Other factors might concern 'materials' or 'manufacture'. Each factor is really a group of related concerns which might be vital to the design or they might be marginal in their relevance. Designers need to establish the relative importance of these factors and generate proposals which seem to offer a suitable compromise.

Partly, designing is this process of seeking suitable compromise. This block examines the tools and procedures which can assist this difficult process. Seeking compromise can be hard enough in relatively simple design jobs such as planning and redecorating a bedroom. When the task concerns the design of a new building, aeroplane, car, or even modest consumer products such as an electric iron, then the process of seeking compromise calls for a wide range of skills, knowledge, abilities and sensitivities.

Design problems usually have many, many possible solutions. One of the main things you will learn in this block is that there is generally *no* simple formula for finding the best design solutions. This is what makes design exciting, challenging, and rewarding. Some people are very good at it and, almost magically, manage to find wonderful new solutions. In this part of Block 2 you will be exposed to some of this magic, and you will be encouraged to tease out some of the systematic knowledge about designing, using an extensive set of illustrations and case studies.

There will be self-assessment questions to help you develop and test what you are learning. Sometimes they will be designed to make you think, rather than testing some specific point in the preceding text. There are no right or wrong answers to many of these questions, since they depend on how you look at things. As you go through the block you should become more confident at answering them. So, try your hand at the following, rather challenging self-assessment questions.

SAQ 1.1 (Learning outcome 1.1)

Look around the room you are in. Write down three or four of the objects you see. Do you think they were designed? Was any part not designed? You might also like to think about the following questions, but do not worry if you cannot answer them. What problems do you think the designer had to solve? Do you think the solutions are good? Do you think the materials used helped solve the problems? Is there any evidence of innovation?

SAQ 1.2 (Learning outcome 1.2)

Let us take a familiar object: a telephone. Mine has a handset plus a base unit with buttons for dialling (note that the term 'dialling' is a hangover from the days when the design featured a rotating dial; see Figure 1.1).

Using any telephone of your own as an example, write a list of those broad factors which you think the design team had to consider when they came up with the particular compromise which is your telephone.

Figure 1.1 Great variation in the design of telephones

1.4 What is good design and good designing?

I stated in Section 1.2 that we are all capable of undertaking design activity. This reiterates what was said in Block 1 about us all being able to engineer to a greater or lesser degree. I am now refining this to the activity of designing. Some people have developed their skills and abilities to a high level and they take part in *professional* designing. The word 'professional' doesn't imply that such people have necessarily elevated themselves to some higher plane, just that they undertake design as their profession. It was also pointed out that the results of design activity continually surround us in the products which we wear, use, ride in, read and perhaps even eat. Judgements about design quality – whether they concern the products or the processes – depend very much on the *context*. Self-evident truths in one context may be irrelevant or false in another. However, in trying to define quality in design it becomes clear that there are common features and recurring themes that are found in many different contexts.

How can you characterize *good* design? A good design is one that appropriately answers a requirement, or meets a stated need. Good design also concerns the anticipation of what people may want. Of course, values come into play here and *wants* may range from the apparently trivial to fundamental needs. Good design might also concern the skilful use of technology, such as materials or manufacturing techniques. It might also imply the exploitation of knowledge, for example information on human size or human perception so as to make a product easier to use.

To be involved in designing, then, means to be involved with using skills (e.g. researching, making, testing), using knowledge (e.g. about things, people, principles), using abilities (e.g. time planning, management), and using sensitivities (e.g. to values, context, markets). It is notoriously difficult to take this further and attempt to define a formula for a process which will lead to successful design. Having said this, in Section 3 I shall present some models of the design process as others have seen it. I hope that you will view these critically and be able to see the strengths and weaknesses of each.

1.5 Designing as model-making and model-using

Any attempt to integrate skills, knowledge, abilities and sensitivities in formulating a design is going to be difficult, and the outcome from one designer's work is likely to differ from another person's attempt. Differences might occur in the unravelling or interpretation of various problems, the generation of ideas intended to overcome these problems or the quality of communication provided to convince others of the quality of your work. Figure 1.1 showed various designs for a telephone. Differences might not only arise in the form of the product, though. There may also be differences in the physical or scientific principles that different designs exploit. Figure 1.2, for example, shows a conventional light bulb and a light-emitting diode, both of which emit light but using completely different physical processes.

In formulating a design, designers use their mental and physical tools for a process which has been termed *modelling*. *Models* and *modelling* encompass a wide range of applications. You may be most familiar with its use to describe a new car (as in 'the latest model') or as a title for those men and women who display the latest fashion creations (fashion models). These popular uses of the term have some things in common with the way designers use it but their meanings have a number of levels which I want to explore here.

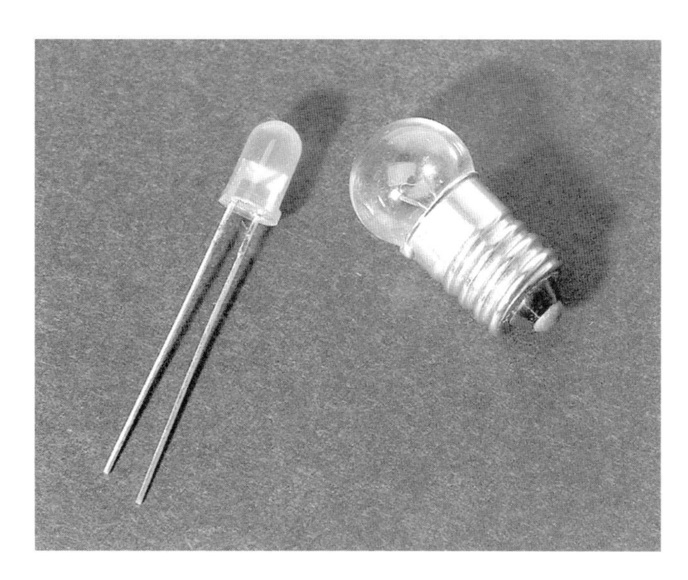

Figure 1.2
A conventional light bulb, which works by heating of a tungsten filament, and a light-emitting diode (LED), which works by light emitted from a semiconducting junction

You will probably be familiar with the use of the term *model* to describe constructions such as an architect's scale model of a proposed building, or a bigger-than-life-size model of a new toothbrush used in a shop display. These are three-dimensional models. I would also use the term 'model' to describe drawings and sketches; in this case they are simply two-dimensional models. An important distinction in the use of such models might become immediately obvious to you, namely their function. That is, the primary function for some models is the communication of information, whereas for others it is to act as an aid in exploring and developing ideas. An architect's scale model (Figure 1.3) which has details of the finished structure, and might incorporate windows, walkways and figures to the appropriate scale, is likely to have been made at the end of the design process, or at least at the end of one design stage. Its primary aim would be to communicate detail to others involved in the process, and this might be achieved using photographs of the scale model.

Figure 1.3 An architect's model of a flour mill

Models in the form of drawings might have a number of purposes. At one extreme, engineering drawings are supposed to facilitate unambiguous communication of precise intention. Sketches, on the other hand, are models which have a primary function to assist an individual designer or design team to creatively resolve problems. Few sketches are formally presented in the way the scale model might be. They are often rough, incomplete and ambiguous (Figure 1.4). In as much as they are 'representations' of thought or intention they can assist communication but it is their rough ambiguity which assists creativity.

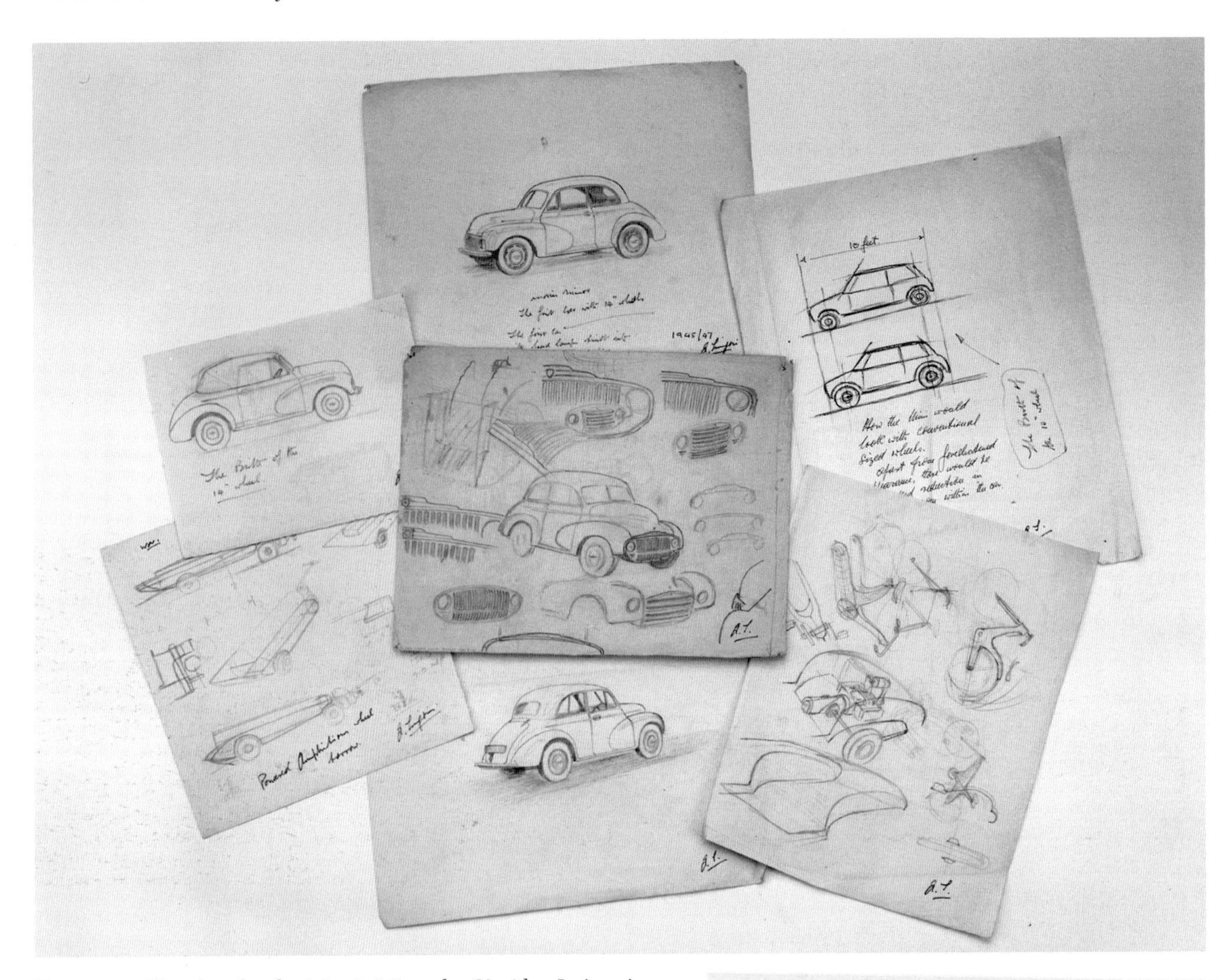

Figure 1.4 Sketches for the Morris Minor by Sir Alec Issigonis

Models might also be expressed as scientific formulae which include the important physical characteristics of the product. So it might, for example, be possible to determine the stresses under load on the gears in a gearbox by a formula including loads, tooth profile and material properties. A computer model of a gearbox might just encode the mathematical formulae but it might also be used to simulate performance where simple formulae do not apply. Computer models might thus take the form of graphic images or, alternatively, highly specified virtual forms.

Another type of model might be those mental pictures we create when we are thinking about a problem or about ways of solving it. We could call this *cognitive modelling*. ('Cognitive' here indicates that this process involves knowledge, experience and reasoning, as well as perception, aesthetics and instinct.)

Figure 1.5 Sir Alec Issigonis (1906–1988), designer of the Morris Minor and the Mini

I could take this one stage further. Consider the skill of designing as, in part, a skill with modelling. Being able to design would imply an ability to link cognitive modelling with various physical modelling tools which might include making cardboard or wooden forms, sketching on paper, constructing plans and elevations as engineering drawings, using engineering formulae, generating computer models, and making electronic circuits on 'breadboards' (temporary circuit boards used for prototyping). As you can see, the design process is a microcosm of the process of *engineering* that we discussed in Block 1.

An ability to make and use models is vital to designing. Models assist in the generation, testing, evaluation, communication and selling of ideas and, as such, their use is not limited to designers. Some modelling techniques are particularly good at supporting several functions at once. For example, some models might be so quick to construct that they can help in the creative generation of ideas. They might also be accurate enough to be used to share ideas with other people involved in the process. The fast sketch drawings created by designer Ian Callum at the conceptual stage of the Aston Martin V12 Vanquish (Figure 1.6(a)) would be a good case in point: such sketches can be assessed for potential appeal in the market before technical questions such as engine power or ignition circuitry are raised (see ▼Ian Callum and the V12 Vanquish▲).

Of course, not all designers are competent with all modelling techniques. Very often designers will specialize. However, a competence at modelling is characteristic of a competence at designing. Engineering design usually involves a lot of modelling, and a lot of it is done on computer. For example, the mechanical parts of a new car or engine are usually modelled in a CAD (computer-aided design) system using a technique called *finite element analysis*.

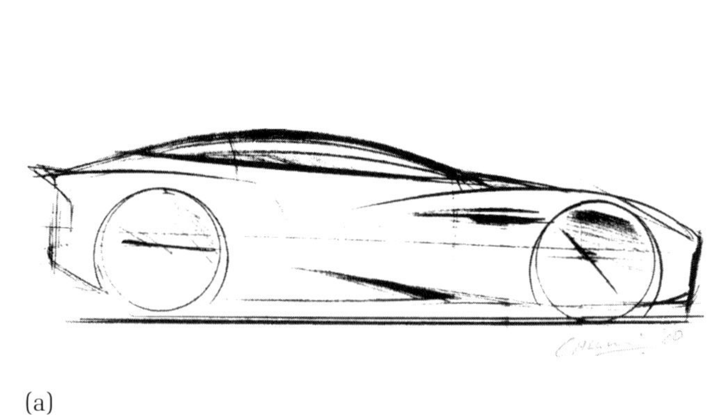

(a)

(b)

Figure 1.6 (a) Sketch drawing used at the conceptual stage of the Aston Martin V12 Vanquish (b) The finished vehicle

▼Ian Callum and the V12 Vanquish▲

Designer Ian Callum, originally from Dumfries in Scotland, studied at the Glasgow School of Art and at the Royal College of Art in London. Early work for Ford led to appointments in Britain, Japan, USA, Australia and Italy, among other countries. Since 1990 he has been commissioned to design for several manufacturers, including Mazda, Range Rover, Volvo, Nissan and Aston Martin.

For Aston Martin, Callum designed the DB7 model (unveiled in 1993). He then designed the concept for the V12 Vanquish (Figure 1.6). This concept was unveiled at various motor shows during 1998, and its favourable reception led Aston Martin to decide to put the model into production as soon as possible.

The main body of the V12 Vanquish is formed from extruded aluminium sections, bonded and riveted around a central transmission tunnel made from carbon fibre. At the front of the vehicle, a steel, aluminium and carbon fibre subframe carries the engine, transmission and front suspension. Engine control, transmission and braking all make extensive use of microelectronics.

The Aston Martin V12 Vanquish was launched in 2001, and is made at the Aston Martin plant in Newport Pagnell, Buckinghamshire. Each car is hand built, although computer-controlled processes are used for the composite sections.

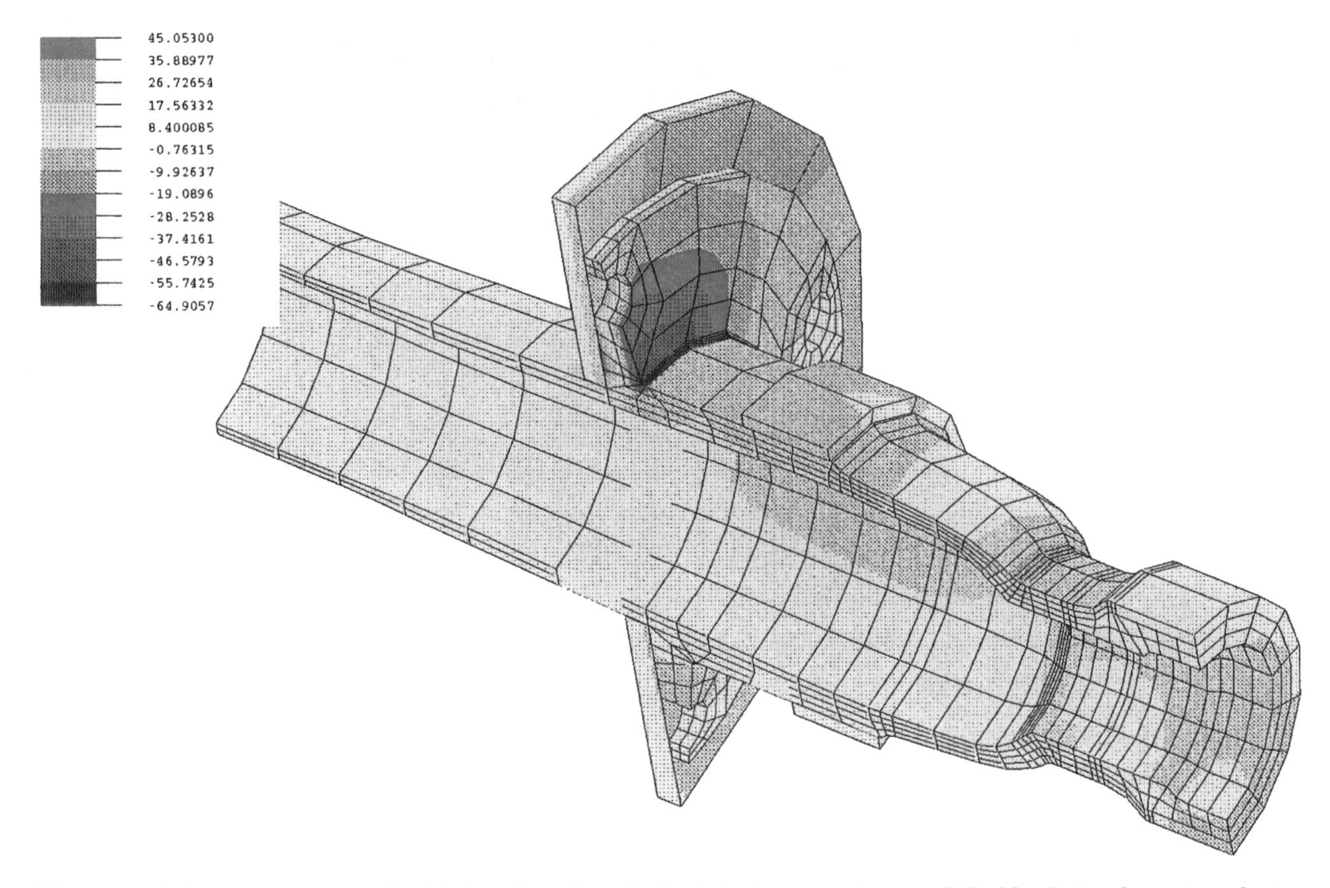

Figure 1.7 Indy racing car rear-wheel hub and section of wheel during cornering, modelled by finite element analysis. The original image uses colour to show regions of stress, as classified in the key alongside the image. The highest compressive stress is in the dark region above the hub

These systems can predict the physical behaviour of the object, including points where it is highly stressed (Figure 1.7).

The next section returns to the issue of *needs*, the exploration of which also involves the exploitation of modelling.

1.6 Design and needs

In the opening sections I raised the question of 'need' as a foundation for design activity. A dilemma came to light: although we might look for needs before generating design proposals, we recognize that some needs may only come to light as a result of getting involved in the generation and testing of various ideas. That is to say, although we attempt to specify our task, define the perimeters to our work, and impose guidelines for planning and direction, it is likely that in many situations we only really uncover detail about the needs when we become involved in designing. Many of the activities we might see in creative problem-solving in design, such as sketching, making models and prototypes and generating life-like computer images, actually have a very important role in problem-finding. Often the real problems can be uncovered only by generating models which provide feedback.

In some instances the feedback required is very particular. In the design of a gearbox, for example, the need might be explicitly stated and the models used in the generation and testing of the proposal can be quite abstract. They might exist only on computer and have no tangible form at all. On the other hand, the design of a new mobile phone might require extensive market testing using life-like models of several versions of the phone in order to elicit the impressions of the kind of people who might be future buyers of the product. Of course there will be many other types of models used to test other aspects of functionality. However, my point here is that vital information about the need might only emerge once a proposal has emerged. The phone manufacturer may find out that some designs do not convey the status expected: perhaps they are too big or too small; perhaps they just look wrong. The remedy might be straightforward, requiring only a few styling changes. Other user feedback might imply a more significant re-think of the whole project. There are two lessons here:

1 Designers have to formulate problems and needs as best they can at the outset of design activity, and yet sometimes they cannot completely define problems and needs before embarking on creative design activities.

2 Designing means getting the correct type and quality of feedback at the correct time. It is vital that the types of models used at any particular stage in the design process be appropriate to the requirements of the design team. With pressures on manufacturers to reduce the time it takes for a new product to be developed and put on sale (Figure 1.8) then strategies for generating the appropriate feedback early in the design process are critical.

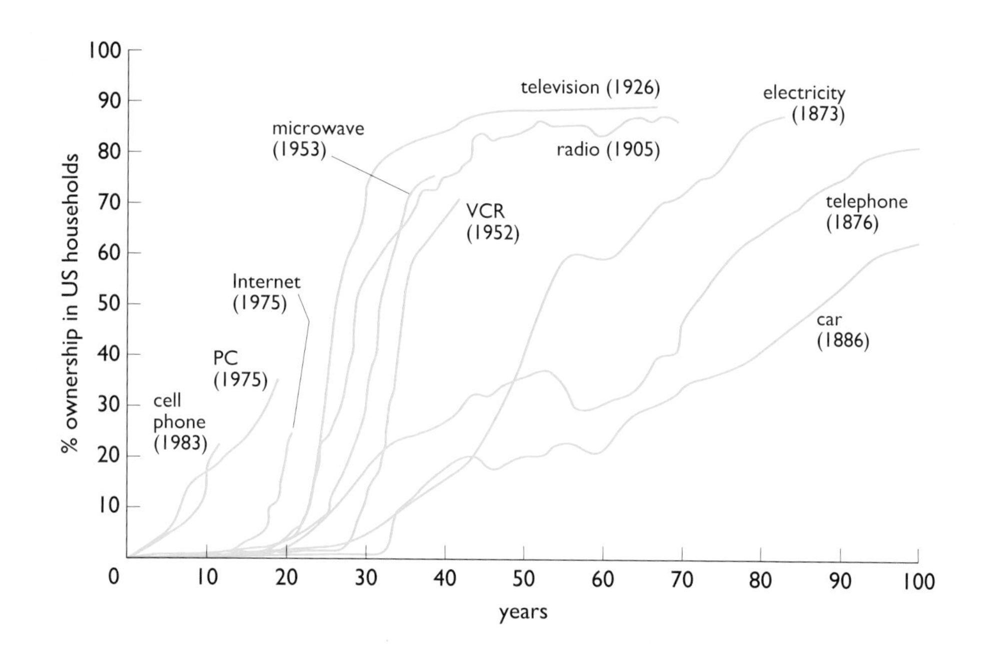

Figure 1.8 Decreasing time taken for new technologies to reach the market

SAQ 1.3 (Learning outcome 1.3)

If you were a professional designer needing to assess the performance, operation or appeal of the products listed below, what sort of models, physical or otherwise, might you use? (*Hint:* think about possibilities like scale drawings, mock-ups, computer simulations, etc.)

(a) A new steel aerosol-spray can to re-market an existing deodorant.

(b) A new family saloon car.

(c) A swim suit.

(d) An elevated section of road near to a town centre.

1.7 Designing as heuristic problem-solving

Generally, solving design problems is different from solving puzzles or mathematical equations. One major difference is that design problems are often not well-specified. This is discussed in more detail in the next section, but one of the primary reasons is the complexity of factors which you began to explore in SAQ 1.2. It means that the properties of the object that the designer is supposed to produce are often not very clear; if they were entirely specified, the designer's job would almost have been done!

Another difference is that design problems don't usually have a single 'correct' solution. Generally there are many possibilities. Sometimes if the requirements are over-specified there may be no solutions. For example, 'Design an aeroplane for 150 passengers that will cost less than £100,' or 'Design a car which will give complete protection to its passengers in any high-speed accident.'

Consider an under-specified problem, such as 'Design me a house for two adults and three children costing less than five times my salary.' There could be millions of possible designs meeting this specification; and there would be millions more designs not meeting the specification. How can the designer find a good design from all these millions of possibilities? One approach would be to try looking at them all, and judging which is the best. This is, of course, totally impractical. In general there is no formula which leads to a good solution, and designers have evolved *heuristics* for solving their problems. A heuristic is a rule or procedure which works most of the time, but sometimes fails. For example, one of the best ways to predict the weather tomorrow is to say it will be same as today. Of course this weather prediction heuristic fails sometimes: you may judge for yourself whether or not it has a pretty good record.

One heuristic used in solving problems with many possible solutions is to reduce the number of options. So the architect designing my house might ask me a few questions, such as whether I would like it to be made of brick, or how many bedrooms I would like, and so on. Each answer would reduce the set of possibilities dramatically. By asking the right questions, the architect could rapidly weed out the majority of 'non-solutions', and begin to investigate some of the remaining possibilities.

Another heuristic used by designers is to look at previous designs to see if there is already a solution to the problem. Often they find solutions to similar problems, which can be adapted to the design problem in hand.

1.8 Design as finding a good problem–solution pair

Design is an interesting form of problem-solving, since part of the problem is to find an accurate way of expressing what the problem is. This may sound paradoxical, but it is quite simple really. For example, if you commission an architect, you generally do not know what is possible within your budget. Some of your desires may be impossible dreams while some may be absolute necessities. Part of the architect's job is to help you, as a lay person, understand these things, explain what is possible, and to help you establish your priorities. Thus you might start the process by requesting a house with five *en-suite* bedrooms, a double garage, swimming pool and tennis court. For most of us, it would not take long to be convinced that the last two constraints make it impossible to find a solution within budget. So, they are removed from the specification, and the problem has changed. If however, you are a *very* keen tennis player you might be prepared to trade-off space in the house in order to build a tennis court in the garden. In this way you can consider the design process as partly looking for the problem (a specification that you can accept) to which a solution can be found. The design then reflects this problem–solution pair.

SAQ 1.4 (Learning outcome 1.4)

(a) Which of the following design problems have a single, unique solution?

 (i) designing a house for a client

 (ii) designing a ball gown for a princess

 (iii) designing a bracket to support a shelf

 (iv) designing a six-lane road bridge to cross the river Severn

 (v) designing a railway locomotive.

(b) Which of the above involve finding a problem–solution pair?

1.9 Design, creativity, invention and innovation

Before proceeding further, it is worthwhile clarifying some of the terminology which surrounds the design process. The words given in the heading above are often used almost synonymously, and in this block we will try to be more specific.

Creativity is the ability to generate novel ideas.

To invent is the process of transforming a novel idea into reality, giving it a form such as a description, sketch or model for a new product, process or system.

An invention is a novel idea that has been transformed into reality and given a physical form such as a description, sketch or model conveying the essential principles of a new product, process or system.

To design is the process of converting generalized ideas and concepts into specific plans/drawings etc., which can enable the manufacture of products, processes or systems.

A design comprises specific plans, drawings and instructions to enable the manufacture of products, processes or systems. A design can also be a particular physical embodiment of a product or device.

To innovate is the process of translating an idea or invention into a new product, process or system on the market or in social use.

An innovation is a novel product, process or system at the point of first commercial introduction or use.

Although *invention* can be the starting point for designing, a study of design is less about invention and more about *innovation* and the *innovation process* from invention to acceptance among users (and competitors). Few products are radical departures from the norm. That is to say, most products belong to a family of similar products. Thus there may be hundreds of different makes and models of digital camera on the market but they broadly share the same technology. Some of the differences may be no more than styling changes to colour or form. Others may typify *incremental innovation* – a process of making small improvements over time. Most of the products of our mass manufacturing culture are variations or *variant designs* based on the same *radical innovation*.

There have been numerous books written about design and innovation. Many have attempted to demystify the process and to demonstrate something typical via studies of various successful innovations. However, there may be a significant flaw in this strategy. The point about successful innovations is that they are atypical. Very few innovations go on to become commercially successful. The vast majority fail, and so to study only the successful ones may not tell us much about the vast majority of innovation taking place. Be cautious if you plan to do any additional reading about innovation and innovators. We are encouraged to believe that successful innovations are the result of some special process or the application of processes by a special individual: the myth of the 'hero innovator'. An alternative viewpoint would suggest that innovation inevitably occurs when certain conditions prevail. However, it is somehow unsatisfying to say that jet engines were inevitably going to be developed in the 1940s because the world was full of innovators and the time was right, rather than to say that 'our' genius, Frank Whittle, invented the jet engine[1]. Either we are all innovators, to a greater or lesser extent, and innovations are common events, usually failures, or there are great innovators who will succeed against all odds, time and time again, and by studying these special people we can hope to emulate them.

1.10 Design is ...

Most design is routine: it's a job. It's people at drawing boards, working at computers, building models, arguing in meetings and learning by doing design work. The subject of design is broad and it takes place in all sectors of industrial society. Although there will be obvious differences between the knowledge and outputs of designers in the various sectors, there will also be considerable similarities in the way they design, the skills they have and the tools they use. It is for this reason that the examples used can tell us much about design and designing as it is found in many contexts. Today much designing takes place via managed groups of people rather than being the responsibility of one individual. The design and development of most new products is just too important, too costly, too urgent and too complex for one person to manage. Collaboration in design may take place between departments within an organization e.g. marketing, production, and engineering, and it can be seen between organizations looking to pool expertise and share outcomes.

As the preceding subsections have shown, there are different ways of looking at design. They are all perfectly sensible in their own ways, and they more or less fit together to give a coherent picture. Does it help an engineering designer to know the wider theoretical background to the discipline? Yes, for many reasons. Engineering designers work with many other kinds of designers, and can benefit from knowing that different areas of design have different traditions and emphases; a knowledge of design theory can also help to identify flaws in a design process. In the rest of this part of Block 2 you will see many examples of this, and hopefully get a good understanding of the phenomena we know as design and innovation.

[1] For a good series of thumbnail sketches of innovators, see Maccoby (1991).

2 Design and innovation 1: the plastic kettle

2.1 Issues of supply and demand

Section 1 attempted to tease apart the various factors and processes that might be found in design activity. Designs can be the result of quite complex interactions which in turn are influenced by context. This complexity is important but I don't want it to be confusing, so in this section I will focus on one example: that of plastic kettles.

In Section 1 I looked at the thorny problem of 'needs', and how as designers we might devise ways of understanding them. Do you think there was a need for a plastic kettle before the first one was introduced in 1978? I cannot imagine somebody making a cup of tea with a conventional metal-bodied electric kettle, and grumbling that what they really 'needed' was a plastic kettle. However, for some reason the introduction of the plastic kettle had a startling effect on the kettle market. Within the first few years of production by several manufacturers, plastic kettles gained 30% of the market share. This raises some very interesting questions which are central to this block. Firstly, might a significant number of users have harboured unspoken complaints about their existing methods of boiling water, including the use of plated and stainless steel electric kettles? Perhaps the market was ready for innovation but it was unable to articulate this because no one had any experience of an alternative to metal-bodied kettles. Might there be some other influence at play: perhaps a movement away from gas towards electric kitchens, which might create an environment ready for innovation? Is it possible that the supply of plastic kettles could have generated a new demand?

2.2 Who dares wins?

As sales were so buoyant, does this indicate that the new plastic kettles were an immediate runaway success? Well, actually no. The early plastic kettles quickly became shabby, primarily because the polymers used absorbed traces of fats present in a kitchen environment and were easily discoloured by sunlight. Also, the surfaces were easily scratched and the fashion-driven colours dated quickly. Even worse, the polymer tainted the water during boiling, giving it a distinctive taste. The first mass-produced polymer kettle was launched in 1978 by Russell Hobbs. It was called the Futura and it was a failure (Figure 1.9). Not only did it suffer from the disadvantages stated above, but it also took a long time to boil owing to its small heating element, which was demanded by the need to minimize the potential fire risk.

Figure 1.9 The Futura – the first plastic kettle

There were technical problems. The Futura was expensive to produce and because it had no lid (it had to be filled via the spout) users were suspicious of the cleanliness inside the kettle[2]. It was killed off because a high number of kettles were returned to the manufacturer. Furthermore, in the company's culture, polymers then became associated with failure, which hindered any subsequent in-house innovation with plastics. It is interesting to note that this design takes its form and general arrangement from the stainless steel kettles with which it was in competition; the innovation was in the use of a new material rather than the shape. It was not until after this that another company launched the first plastic 'jug'-shaped kettle which did so much to invigorate the kettle industry.

Although the 'demand' side of the supply-and-demand picture was problematic at this time, radical changes can be observed in the supply of plastic kettles. The increasing availability of cheap plastic mouldings together with the necessary heating elements, switches and cords meant that it was possible for many new companies to set up in business assembling and marketing plastic kettles. The significant investment required for the fabrication of stainless steel kettles did not apply to these new assemblers, and the number of companies producing plastic kettles increased tenfold. They joined established companies such as Russell Hobbs who had their own in-house design, manufacture and marketing capability. Figure 1.10 illustrates the expensive stages involved in the production of a stainless steel kettle; Figure 1.11 shows part of the production line at Russell Hobbs where metal is stamped, cut, bent, soldered and polished: processes that require a skilled workforce.

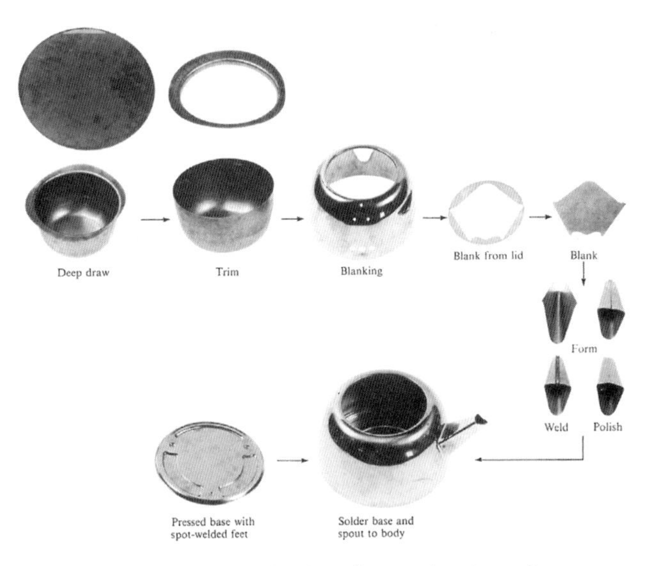

Figure 1.10 Stages involved in the production of a stainless steel kettle

Figure 1.11 The mass production of stainless steel kettles

By comparison the production of a plastic kettle body is an unskilled process. One worker takes the plastic kettle body from an injection moulding machine (into which molten plastic is forced into a mould under pressure, and where it cools and solidifies into the desired shape), snaps off the sprues (residual projections resulting from solidification of plastic within the mould's feed channels), and puts the kettle body in a bag (Figure 1.12).

Figure 1.12 Machine-minding in the injection moulding industry

[2] Known in the trade as the 'dead-rat' syndrome: if you can't see inside the kettle, how do you know there isn't a dead rat in it?

After an enthusiastic welcome for the innovation, the market displayed some resistance to plastic kettles. They were not perceived as good value for money, and word soon spread. However, the supply was rapidly increasing, thus driving the price of plastic kettles down. It was, and still is, very convenient to be able to boil water safely and quickly and the manufacturers knew there was a huge market if they could overcome the problems. Materials science soon solved the problem of the unpleasant tainting of the water during boiling, and as the market grew so the price of plastic kettles tumbled. Consumers were faced with some stark realities. They could purchase a metal kettle, which would undoubtedly have a long life, but at a considerably higher initial price than the plastic kettle. Alternatively, they could buy a new plastic kettle, add the status of a modern bright object to their home and replace it, if necessary, in a few years because of the cheap price. In spite of some companies continuing to promote the stainless steel kettle (see Figure 1.13) the plastic kettle won out.

Figure 1.13 Advertisement for the Russell Hobbs stainless steel kettle

Sales of plastic kettles continued to increase and manufacturers invested in the victor in order to secure for themselves a share of this lucrative market. New shapes, colours, surface finishes and styles of applied decoration appeared, but the real influence on demand seems to have been the shaping of the kettle body in the form of a jug.

It is not at all certain that anybody really benefits by changing the kettle shape from a pot to a jug but it has become the standard form for electric kettles today. Some advocates point to the ability of jug kettles to boil smaller quantities of water and thus to address the needs of those users who want to make one or two hot drinks at a time. This is achieved because the tall, thin jug kettle requires less water to completely cover its element. Clearly this has potential benefits for whoever pays the electricity bill and for the environment, but research has shown that most users continue to put as much water into a jug kettle as into a stainless-steel kettle!

The first successful plastic kettle was jug-shaped. It was designed by a consultancy called Action Design in 1979 and was produced by Redring, a company experienced in manufacturing the heating elements but new to the kettle market. In 1981 the Redring jug kettle saw more than a quarter of a million sales in Britain and together with export orders, contributed £16 million to company turnover. As with other contemporary plastic kettles the material and technical components were not ideal but its success in the market was observed by other manufacturers and it inspired the search for better polymers and technical improvements.

Exercise 1.1

How did the first polymer kettle compare to existing metal kettles in terms of:

(a) looks,

(b) cost,

(c) boiling time,

(d) taste of the water,

(e) durability.

2.3 The significance of 'need'

I started this section with a reference to needs and I want to return to this now. There was no explicitly stated 'need' for a kettle made of plastic or a kettle shaped like a jug. Even if a manufacturer had undertaken comprehensive research using questionnaires, interviews, or brainstorming sessions I doubt whether it would have come up with a brief for a plastic jug kettle. My point is that successful innovation is not necessarily directed by needs. The innovative plastic kettle was a result of:

- the availability of new polymer materials;
- the development of techniques for forming these new polymers;
- the emergence of cheap manufacturing capacity in the UK and overseas;
- the growth of the component industry;
- changes in retailing which made cheap consumer products widely available;
- growing consumer affluence which allowed increased spending on the home and domestic products;
- changes in attitude to the material culture which enabled people to consider previously valuable household tools as disposable items.

Yes, the plastic kettle was partly concerned with needs: the need to boil water safely and cheaply; the need for status associated with modern consumer products; and the need for novelty; but it would be wrong to view innovation as merely a response to market needs. Given the above conditions, if it was possible to produce a working plastic kettle, someone was going to try it. The style of the jug kettle was key to the success of the plastic kettle, but we should not overlook the fact that the technical issues associated with its manufacture in plastic and its jug form were difficult and complex.

Another lesson is that functionally superior products will not necessarily win in any given market. It is also salutary to observe that being first in a new market is a perilous business. Often it is not the pioneers that posterity remembers, but the people who came afterwards and who are probably the first to enjoy commercial success.

Exercise 1.2

You are going to undertake some simple product analysis.

Make two lists on a piece of paper in two vertical columns. With reference to a plastic kettle that you are familiar with, write down as many of its good points or qualities as you can think of in one column. In the other column write down all the weaknesses or faults of which you have become aware. These are good and bad points as you have experienced them: you needn't try to think of the issues concerning manufacture or marketing. You may want to use some of the positive and negative points I have raised in the discussion in order to start you off.

3 Models of the design process

3.1 Reprise on models

As discussed in Section 1, models – physical or conceptual – are used extensively in design to give information about what the final product might look like or what its properties will be. This is a way of trying to make the design more understandable during its initial stages. In the same way, many people have produced models of the design process itself, to try to understand better the optimum route to producing 'good' designs. In general, models are used to represent things for some purpose. In Section 1, I discussed drawings and constructions as models. One thing all models have in common is that they are incomplete in one or more respects when compared to the thing they represent. Models are used to explore some properties of things; other properties considered to be unimportant for the purpose in hand may be excluded from the model. Thus the weather map seen on television each day is a kind of model. It is not a full and complete picture of the weather; it is a simplified version that enables us to understand the important information quickly. The map of the London Underground is another famous example of a model.

The models that will be considered in this section illuminate various aspects of design. Each has its own advantages, and each has its own shortcomings. We can ask also if these models are useful to designers, and if so, in what ways. This will be discussed at the end of this section.

3.2 Building a simple model of design

In this section I am going to make a model of the design process. I'm not going to be specific about what is being designed; I want the discussion to be very general in a way that will apply to many products.

Design is a process with a beginning when the decision is made to design something; and an end when the design is complete, and the designed object is fabricated. So let our first model be Figure 1.14.

Figure 1.14 A first model of design

This model does not say much and is not very useful. After all, what is being designed? At the very least we need some kind of specification. So let Figure 1.15 be the next model.

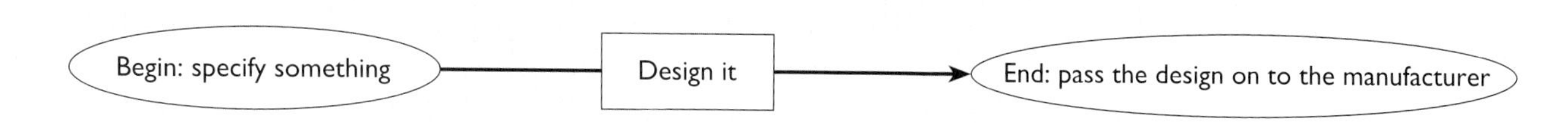

Figure 1.15 A second model of design

But what does 'Design it' in Figure 1.15 mean? Originally there is nothing but the specification, and from this a design is *generated*. Is the design good or bad? In order to judge this, the design must be *evaluated*. So now the model becomes Figure 1.16.

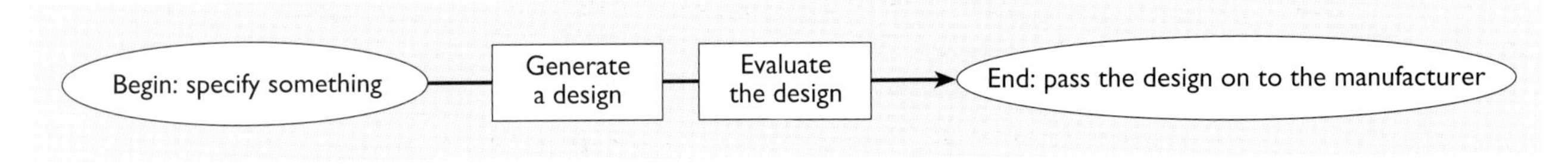

Figure 1.16 A third model of design

Now, suppose the evaluation of the design suggests that it did not meet the specification very well, or that the object would not work as intended. What happens then? Usually it's 'back to the drawing board' – or possibly back to the specification in order to make changes to it. So we add an *iterative loop* into the model to get a *design cycle*, as shown in Figure 1.17.

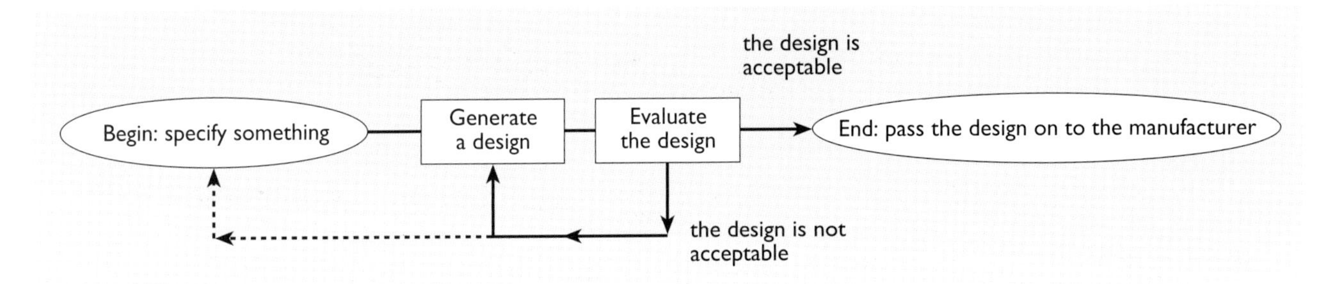

Figure 1.17 A fourth model of design

If the design is judged to be unsatisfactory, there is a chance to have another go. The designer or the design team might even judge that the problem lay in the initial specification, and so could return and amend it before starting again or making changes.

For the purpose of this discussion, and for the moment, this model will be considered satisfactory.

SAQ 1.5 (Learning outcome 1.5)

We have just 'designed' a model of the design process itself. Suppose the specification was 'to produce a model of the design process'.

(a) How many generation stages were there?

(b) How many evaluation stages were there? What was the decision at each stage?

I hope that as this model was constructed, you understood the steps involved in each development stage. If you were being critical, you may have anticipated the problems within each stage, and you should anyway have been making your own judgements as the model was developed. Perhaps you felt that the development should have been different, and perhaps you felt the final evaluation was rather complacent. Perhaps you thought that we should be able to produce a better model than that?

One of the important features of model-building is that, as with design itself, everyone can have a go, and the resulting models reflect personal tastes, knowledge and interests. Other models will be discussed in this section, and they are other people's attempts at representing the design process as they see it.

As you view these models you should be *critical*. Ask yourself what is *good* about them, and what is *bad* about them. In other words, *evaluate* them. Your criticisms should be constructive. Criticisms such as 'This is rubbish' do not take things far forward, either in developing a better model or in developing a better design. It's much more constructive to say: 'This model fails in such and such a respect because of this or that reason, and this problem could be overcome by doing the following.' In other words, constructive criticism involves identifying weakness (evaluation) and suggesting new ways of overcoming them (generation), and it is an essential part of design.

One misleading aspect of the diagrams that are used to model design is the suggestion that, when designers go round the generate–evaluate loop, they go back to where they started. Of course they don't, because by going round the loop they learn. Thus our simple design model could show the process unfolding in time, so that a spiral replaces the loop (Figure 1.18).

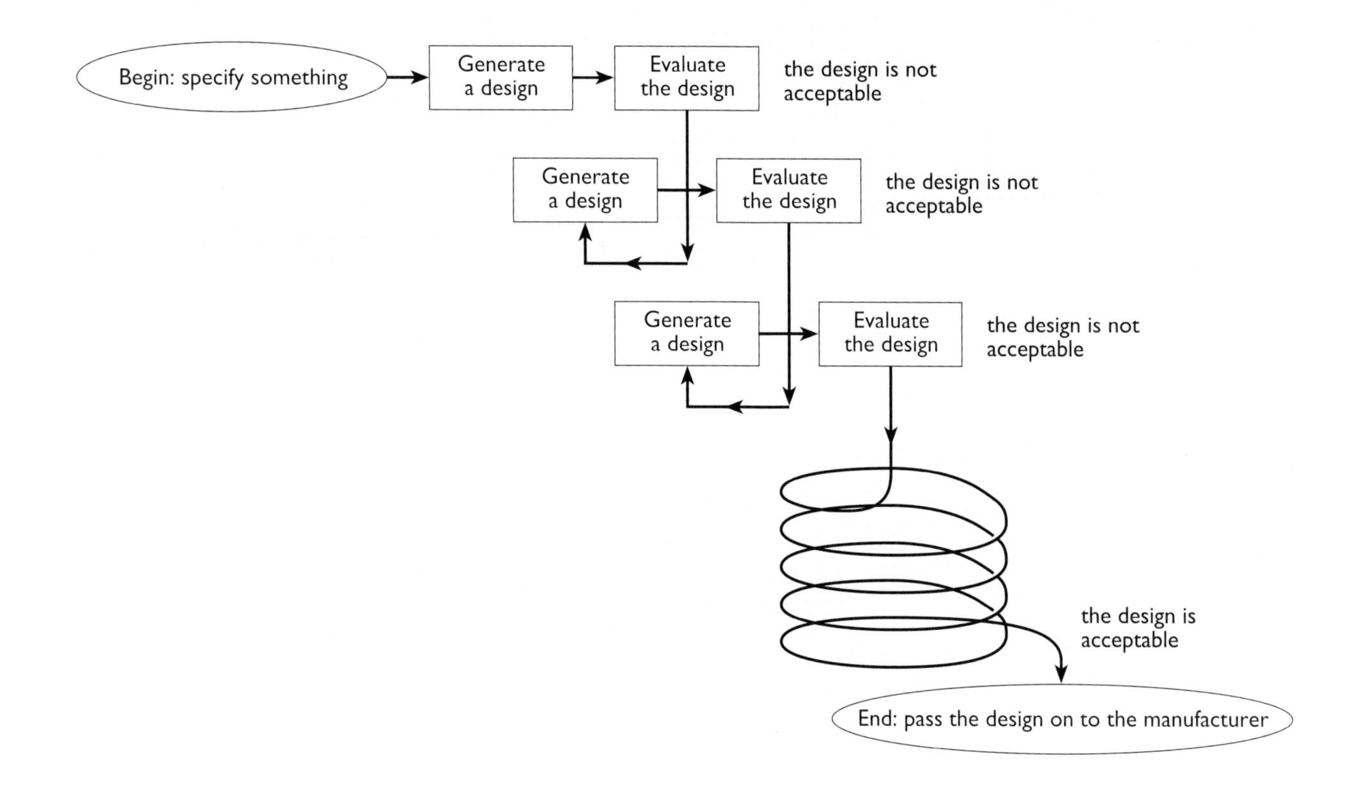

Figure 1.18 A spiral model of the design process

Is this now a useful model of the design process? Let's consider this by applying it to our example from the previous section: the development of the plastic kettle.

The emergence of the plastic kettle owed more to Russell Hobbs taking advantage of new technological developments and spotting an opportunity for a new product rather than the company meeting real needs of users. However, there were evident needs, and these were to do with low costs, convenience and fashion. The starting point was the design of the Futura kettle by Russell Hobbs. Sales of this kettle were not high, possibly because it had no lid. In this case the market evaluation of the design was critical. The market didn't like it, and the company lost a lot of its confidence for producing such innovative designs.

However, plastic-kettle making did not stop, because someone else learned from the Futura experience, and came up with a different design. This design sold, possibly because it was novel, and possibly people believed it was more energy efficient because of the jug shape.

Other manufacturers observed the success of this product, and new plastic kettle designs evolved, to be evaluated by the market. For the evaluation of the kettle design, we have to add *people* (meaning the general public in the form of consumers) to the diagram. Our earlier spiral diagrams did not say who was doing what. Also, the spiral for the kettle development has no end. Design modifications will continue after the kettle is launched because of feedback from customers.

SAQ 1.6 (Learning outcome 1.5)

Consider the development of the plastic kettle, as discussed above. Construct and label a spiral diagram giving a model of the development of the plastic kettle from the Futura onwards.

3.3 Other models of design

Having now begun to make simple models of the design process, let us consider whether this is useful. Are such models helpful in working on and developing a new design? I will postulate that there are no *practically useful*, general theories of design and that the study of successful designers does *not* necessarily lead to practices that help us to produce designs ourselves. Rather, I take the view that design is strongly situated – what constitutes good design methodology in one context may not be universally true – and that general understanding is built on experience of different contexts. Specific understanding of how, say, your own company understands, manages and executes design is hard won and valuable.

This is not to say that there are no models at a high level of abstraction that inform our understanding of the design process. The models presented in the previous section had useful messages in forming a basic understanding of how designers work.

In the following pages we will examine some of the models proposed by a variety of experts. Then you will have the chance to decide what you think about them, and the possibility of a general and useful model of design.

3.3.1 March's model: philosophical

The act of *synthesis* is central to design. Synthesis means bringing things together to make something new, something different from the constituent parts, something *synthetic*. March's picture of design (Figure 1.19) describes three types of process that act together in order to create a new design.

The process of *production* produces an initial design proposal, from many possibilities, that is a candidate to solve the design problem in hand. The process of *deduction* applies known theories and understanding to predict the performance of a design proposal. The process of *induction* evaluates a design against specification. Resulting changes and refinements help generate a new design proposal (production again). The cycle repeats, taking a designer towards a solution.

These ideas are intuitively attractive. It is easy to imagine all of these processes taking place in some sort of suitable mix. For example:

- Do I build a timber-frame house on my plot of land, or a brick-built house (production of a candidate idea)?
- Some calculations of energy efficiency, cost of labour, and a judgement on the resale value leads me to a conventional UK design in brick and glass (deductions about the idea).
- Whilst planning a conservatory I hit on the idea of using it to channel warm air into the main living space, so refining the design (induction to further design ideas).

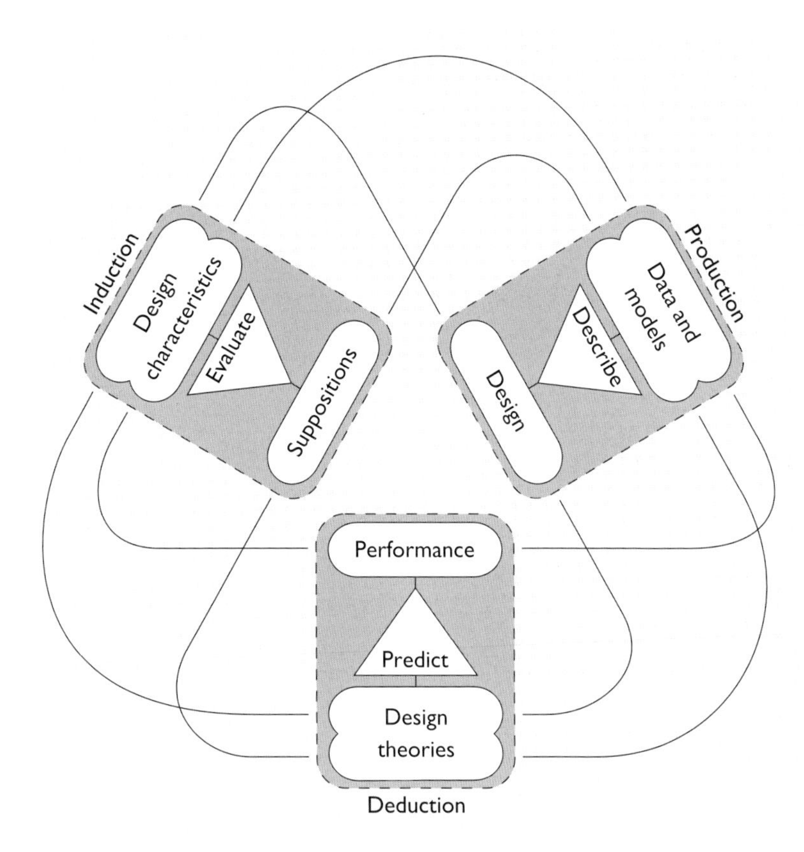

Figure 1.19 March's picture of design. Adapted from March (1984)

The degree to which each component is present in the mix could vary. Designing an electrical circuit with similar characteristics to a previously designed circuit might only involve the deductive process. By contrast, designing an advanced space vehicle would require a great deal of production.

The temptation to break down design into detailed plans is, however, irresistible.

3.3.2 BS 7000 model: practical

There are many more representations of design processes and issues in the academic and industrial standards literature. All are attempts to move down from the broad philosophical view of Figure 1.19 towards the practicalities of the design process. The ones that we will consider all come from engineering design.

By contrast to the rather abstract and elaborate academic model of Figure 1.19, the British Standards BS 7000 picture of design (Figure 1.20) specifies a direct route from start to finish.

SAQ 1.7 (Learning outcome 1.5)

For each of the boxes in Figure 1.20, identify who would be involved in each of the steps of designing and constructing a new house for yourself, and what each of the steps would comprise. (*Note:* some of the boxes will not apply.)

What's missing from the house-building process if we apply this model rigidly?

If you have experience of even modest building schemes, you will know that the architect must be there to supervise the builders, otherwise *production* will be a mess. And you will know that the designer sometimes makes

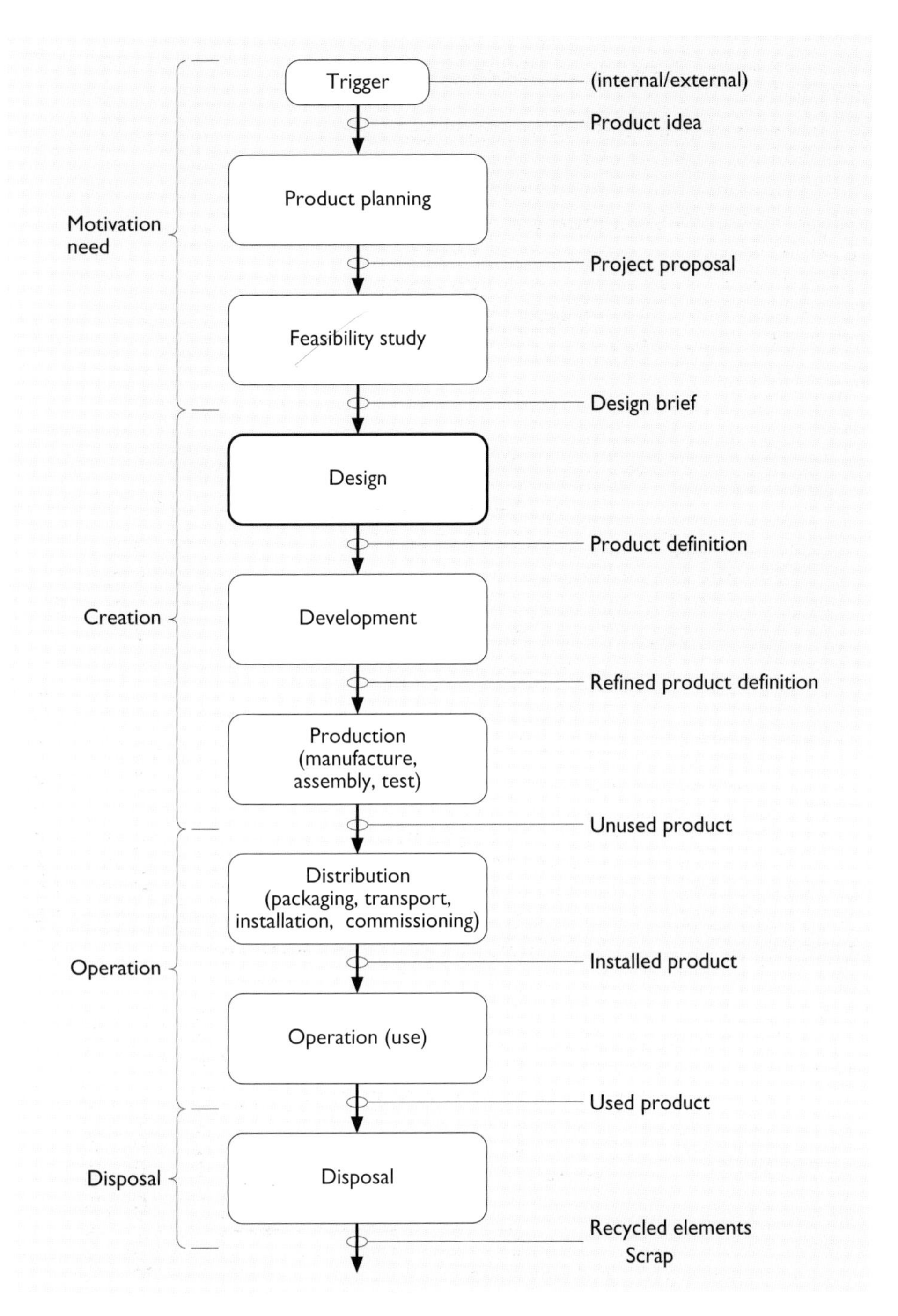

Figure 1.20 The BS 7000 picture of design

mistakes and has to negotiate with the builder to find a solution on site. Thus there may be a loop back from production to design. None of these iterative loops is in the model (and the model contains steps which are not relevant to our example).

The model does not take into account the financial problems you will have as the cost escalates due to the specification changing. And the model does not take into account that you may have to change the specification, even quite late in the process.

SAQ 1.8 (Learning outcome 1.5)

(a) Suppose you were designing a passenger jet aircraft. Go through the activities listed in BS 7000 and try to say what would be going on at each stage during the design of the aircraft.

(b) Do you think it might ever be necessary to go back to review previous steps in the design? Which ones?

(c) Do you think the BS 7000 picture of design would describe the process adequately?

Even though it has the 'gold seal' of the British Standards Institution, this model has its shortcomings. But although one can criticize it, it is not completely wrong or completely useless. At some level it describes reasonably well the flow of activities in the life of developing a product. Some parts of the diagram will fit some products better than others.

So, how can the model be improved?

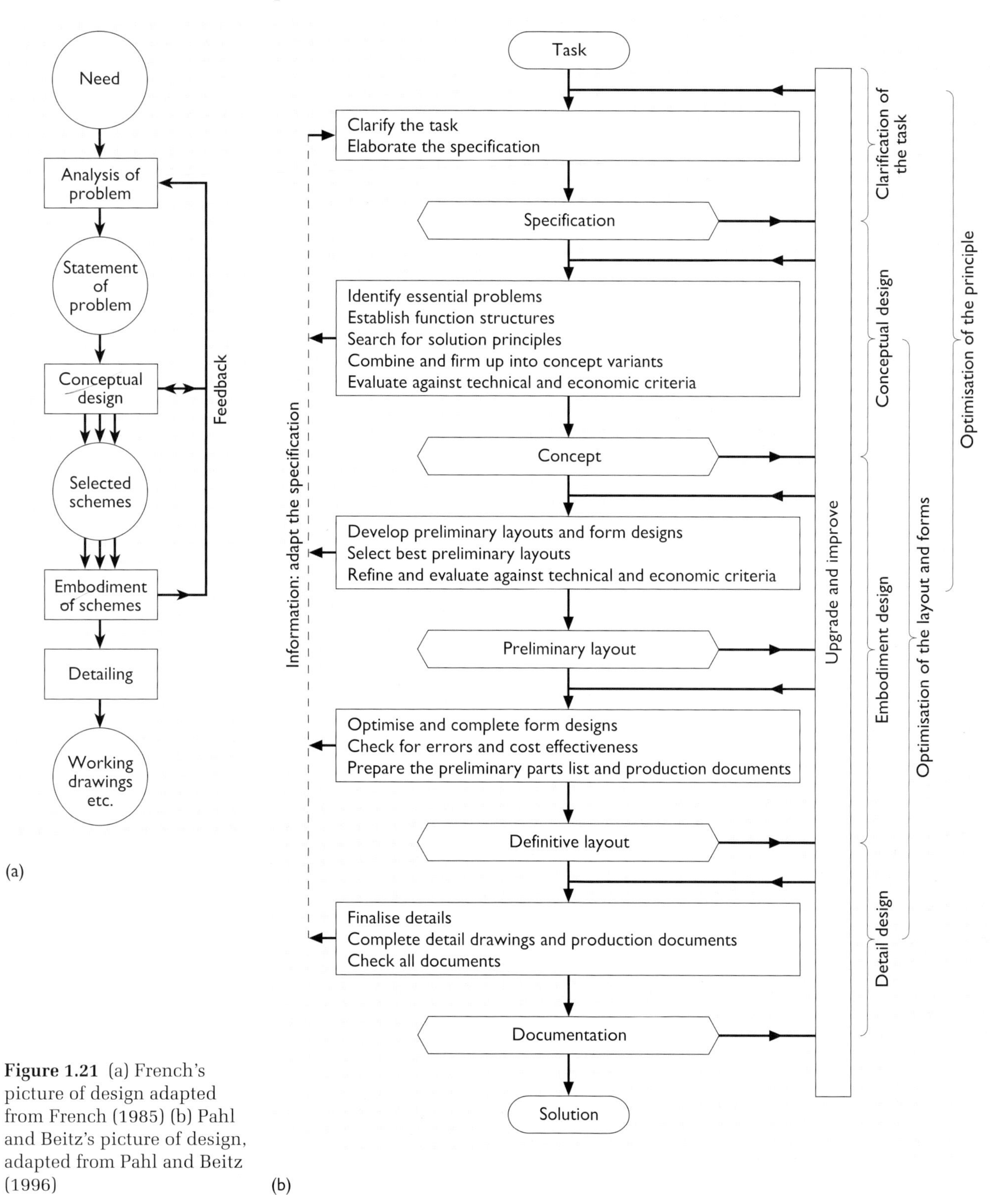

Figure 1.21 (a) French's picture of design adapted from French (1985) (b) Pahl and Beitz's picture of design, adapted from Pahl and Beitz (1996)

French's picture of design (Figure 1.21(a)) starts with a need and proceeds sequentially through formal stages with feedback. French's picture is relatively simple. For example, if you were building a house you would start with a *need* (somewhere to live), you would prepare and analyse a *specification* (the price, number of rooms, location, etc.), and you would *state the problem* to the architect who would come up with some *conceptual designs*. You would *select* those you liked best. It's not clear to me what *embodiment of schemes* means in this context, perhaps an integration of the best features of several schemes and certainly the production of drawings. These drawings would eventually have to have all the *details* specified for the builders, and they would be recorded as *working drawings*. The drawings are also used to *communicate* the design from designer to fabricator.

An important feature in French's model is the possibility of 'going round the design cycle'. At almost every stage, the designer must expect to revisit earlier stages to make changes so that the design 'works' lower down. French's model captures a very important part of the design process, that it is *iterative* in the way it cycles round until a satisfactory solution/scheme is found.

Exercise 1.3

What is the important feature of design captured by French's model which appears to be absent from the BS 7000 model?

French's format is elaborated by Pahl and Beitz's picture (Figure 1.21(b)), which follows the same backbone, but adds more labels, more definition, and more feedback. At first sight this picture is harder to understand than French's.

The Pahl and Beitz model is one of the standard models in the engineering design literature. This model shows how designs are 'worked up' from loosely defined ideas into something much more definite as the process proceeds. At the beginning the designer is, understandably, unsure what the solution will be. As the design proceeds the solution (or solutions) becomes clearer.

Of course, sometimes the chosen solution path may fail; then it's back to the drawing board again round one of those loops. This model also allows the designer to go back and modify decisions made at earlier stages.

Exercise 1.4

Both French's model and that of Pahl and Beitz assume there is a need or a task to be fulfilled. Was this the case in the design of the plastic kettle?

SAQ 1.9 (Learning outcome 1.6)

Use the example of designing a five-bedroom house to answer this question.

In your opinion, which of the French model and the Pahl and Beitz model best represents the design process? To formulate your answer, take a list of the good and bad points of one model, alongside a list of the good and bad points of the other. Then weigh up the pros and cons, and give your considered judgement.

It is interesting to note that by listing good and bad points in my answers to SAQ 1.9, I have convinced myself that none of the models we have studied is perfect. This is an important point. The models exist so that we can try to gain an understanding of how designers work; any model will either miss some key activity or will not be generally applicable. There is no single perfect model of design.

3.4 Conclusion: are models useful for practising designers?

In this section I have:

1 built a simple model of the design process;

2 considered other models of the design process:
the March model
the BS 7000 model
the French Model
the Pahl–Beitz model.

One conclusion that I have reached in this section is that none of the models proposed is a perfect description of the design process for even one specialized area of design. Like all models they capture some aspects of reality, but lose others. Thus the idea of a 'golden road' to successful design embodied in a single authoritative model seems untenable.

In reality, design is undertaken by humans who will reflect continually on the job in hand and on the best way to achieve the desired result. They will not approach their designing by adhering rigidly to one particular model.

Despite the artificiality of the models, however, it is true that most industries must have tight procedures in order to manage the complex and very expensive interaction between their designers and/or between design teams. Companies may impose procedures based on formal design models, simply because they have to have some explicit set of procedures to follow.

Design cannot be managed routinely if it is to result in good outcomes; whatever the procedures adopted, the success of the resulting designs will be unpredictable. Poor designs will be routinely produced, and it is better to identify and eliminate them early than to assume that a 'good' design model will eliminate them.

One thing is certain. Design, as it is currently practised in most industries (including engineering design), is *not* an ▼Algorithm▲. However, it certainly benefits from being well-organized, and designers benefit from knowing that there are stages in the process which can be identified.

All the models we have considered had features which corresponded to some design processes. In my view this makes the models useful. I would even suggest that reflective practitioners might find them useful for comparing with their own design process. If there is something in one model which resonates with practice, and something in another model that resonates with practice, it may be possible to combine these two ideas to give a new, *bespoke*, model of the particular design process.

▼Algorithm▲

An algorithm is a sequence of well-defined operations that lead to the solution of a problem.

That definition, though, doesn't quite capture one of the distinguishing features of an algorithm, which is that the operations used to reach the solution should be specified as straightforward, unambiguous instructions that can be performed in a routine or mechanical way. So, for example, a rule for dividing one fraction by another which said, 'Turn the fraction after the ÷ sign upside down, and change the ÷ to ×.' (*Sciences Good Study Guide* p. 328) is an algorithm. On the other hand, a rule which said that a laboratory report should consist of an introduction, a description of the method, a set of results, a discussion and a conclusion would not be an algorithm, because simply following those rules would not generate the report.

The implementation of an algorithm should not require additional creativity or problem-solving on the part of the person or machine that performs the algorithm. However, devising the algorithm in the first place is usually a highly creative business. A software engineer who writes a computer program uses creative design expertise (and other skills) in order to create an algorithm that can be performed by a machine.

Thus I am suggesting all these models might be a rich repository of design relationships. Familiarity with them may enable designers to disassemble them into pieces, which work 'locally', even though the whole does not work. Then the designer may be able to assemble pieces from the various models, and possibly some pieces they have discovered for themselves, to create their own models which they feel describe their individual design process.

I think most designers do have such models so that at any stage they feel comfortable that they know what they are doing, and why they are doing it in relation to previous and future activities.

The models you have seen in this section are part of what might be called 'the culture of design'. Most designers will have encountered aspects of the models, and those designers who are reflective practitioners will have thought about the relationships between the parts of the process. To this extent, although we don't think that the models are perfect, they inform practitioners about the possibilities they may encounter.

In a typical designerly analysis, we come to the conclusion that the models have some useful features for designers, but they cannot be blueprints for the design process. Our conclusion is that the models are neither useless nor essential.

Having described the models, I now want to examine, in Sections 4 and 5 some important aspects of the process which appear in all of the models, that is, conceptual design and the route from concept to prototype.

4 Conceptual design

4.1 Establishing the design space

In Section 1.1 I proposed that, to some extent, we are all designers. We are all confronted by problems in our daily lives and we all use methods and procedures to overcome these problems. At the most basic level, our design problem might be deciding which clothes to select from our wardrobe or what to purchase when out shopping for groceries. When I'm in this position I think of where I'm going, or what cooking I intend to be doing. In a very simple way I'm trying to define my problem, or at least the 'space' in which my problem lies; that is, my *problem space*, as opposed to the real physical space in which I operate. While trying to solve this problem I might try to imagine what I'm going to look like in various clothes (using cognitive modelling!) or, if I'm buying them, I try them on (physical modelling). Here I'm trying to establish the *solution space*. I might find myself making many iterations of trial and evaluation before I feel comfortable. That is, iterations between the problem space and the solution space until I find that problem–solution pairing I discussed in Section 3. These iterations allow me to broadly develop the *design space*.

We might also come across this process of defining the design space in other fields such as laying out a page on a word processor, designing our garden, or planning to redecorate a room. This is not really designing in the sense of the plastic kettle I discussed in Section 2. There is a difference in complexity between the designing we do in our daily lives and the activities undertaken by professional designers. This section takes a closer look at this complexity and begins to unravel the design activities which are crudely represented in the models of design discussed in Section 3.

There are experts in all of the various different design fields, and for many people designing is their profession and their livelihood. These experts will be people who

- have specialist knowledge;
- are likely to be familiar with the types of problems which can arise in that area;
- have a competence with making and using the types of models used in that field;
- may have a wide range of contacts to assist their work.

When your own designing becomes too complicated you might have found yourself calling in an expert: for example, using an architect to help you achieve a new extension to your house. He or she may be able to assist you with the concept designs or to identify the function restrictions encapsulated in the building regulations.

In the concept or conceptual stage of design, the emphasis for the designer or design team is on defining the appropriate solution space which best matches the known problem space. Of course, as we have seen, achieving this is not easy for a number of reasons. Firstly the problem may not be well defined[3]. Secondly the act of generating solutions can allow us to reconfigure the problem. A spiral of development is created where defining a new boundary to the problem space facilitates even more new ideas in the solution space which further tests the problem space.

[3] There is an old but interesting paper by Rittel and Webber on the unique nature of design problems in Cross *et al.* (1973).

In order to find the appropriate design space, designers usually have to go through iterations of generating ideas and testing them against the known problem. This may result in changes to the formal statement of the requirement (the product design specification) and/or changes to the proposal. At the concept stage there can be huge variations in the specification and in the types of proposal offered but the objective will always be to do two things. Firstly, at the concept stage designers seek to improve the specification so that it more accurately represents what is actually required. Secondly, designers seek to offer a range of ideas which meet, as well as possible, the developing design specification. So concept designing has the tricky function of offering creative interpretations to an emerging problem. To help with this dual activity, designers use a wide range of modelling skills in addition to their knowledge and experience.

I shall develop my discussion of conceptual design, and particularly this notion of solution space, using two examples: the design of the hull of a sailing boat and the development of human-powered flight.

4.2 Conceptual design in sailing boat hulls

Hull shapes for sailing boats typically fall into the three types shown in Figure 1.22. The 'fin and skeg' design is typical of modern mass-produced yachts and small dinghies. The long-keel design is typical of working boats designed to be handled by a small crew. These two forms represent extremes of stability and performance. The dish shaped hull of the fin-and-skeg will skate on the sea's surface and turn quickly, whereas the long-keel yacht will be stable in a rough sea and manoeuvre slowly enough for a small crew to have time to handle sails and ropes. The compromise design is typical of a 1950s yacht. It has better turning performance than the long-keel design and more stability than the fin-and-skeg design.

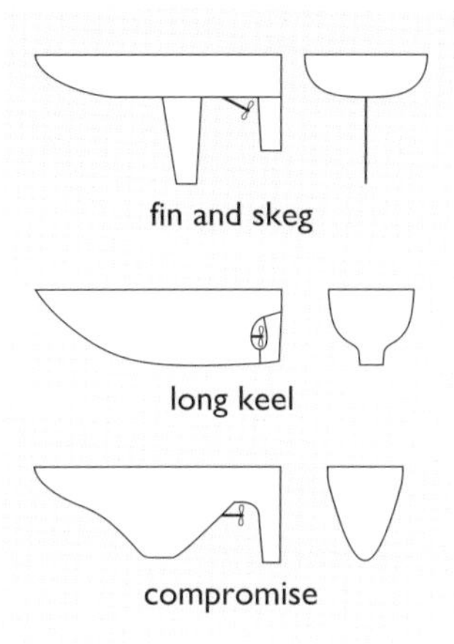

Figure 1.22 Three forms of sailing boat hull

Many large, modern sailing boats, designed and made for both cruising and racing, have the same overall form as small dinghies. These dinghy shapes are cheap to manufacture, lively, quick, and none too comfortable. You might imagine that they are advertised for speed rather than ease and cheapness of manufacture.

The sketch of three hull forms shows no superstructure (cabins etc.) on the hulls. Clearly somewhere must be provided for the crew to eat and sleep, for storage of sails and equipment etc. Depending on the design of the hull of the yacht, there will be different available solution spaces.

There is a clear compromise between sailing performance and interior space that a designer must consider. Anyone sailing on the boat will need to have headroom. This can be achieved either by building up from the deck level of the keel – which introduces surfaces that are exposed to the wind and affects the sailing properties – or by situating the living space inside the keel itself. If a fin-and-skeg design is chosen the only way to provide headroom is to build upwards, whereas a long-keel design can provide headroom inside the hull. Whether the headroom that is required is for standing or sitting extends the problem. A large, high cabin to maximize the interior space and carry people in well-lit comfort is a common design solution for modern pleasure yachts that are unlikely to be used in difficult conditions.

Consider now, how space might be subdivided in the two extremes of fin-and-skeg design and long-keel design. A decision tree can be created to show the different design solutions, Figure 1.23. (Note that for simplicity the fin-and-skeg design is shown simply as a finned keel.)

The design decisions concerning how space is apportioned as the design progresses are shown as a hierarchy. Clearly, the choice between a shallow hull with a fin keel and a deep hull with a long keel is a high-level decision

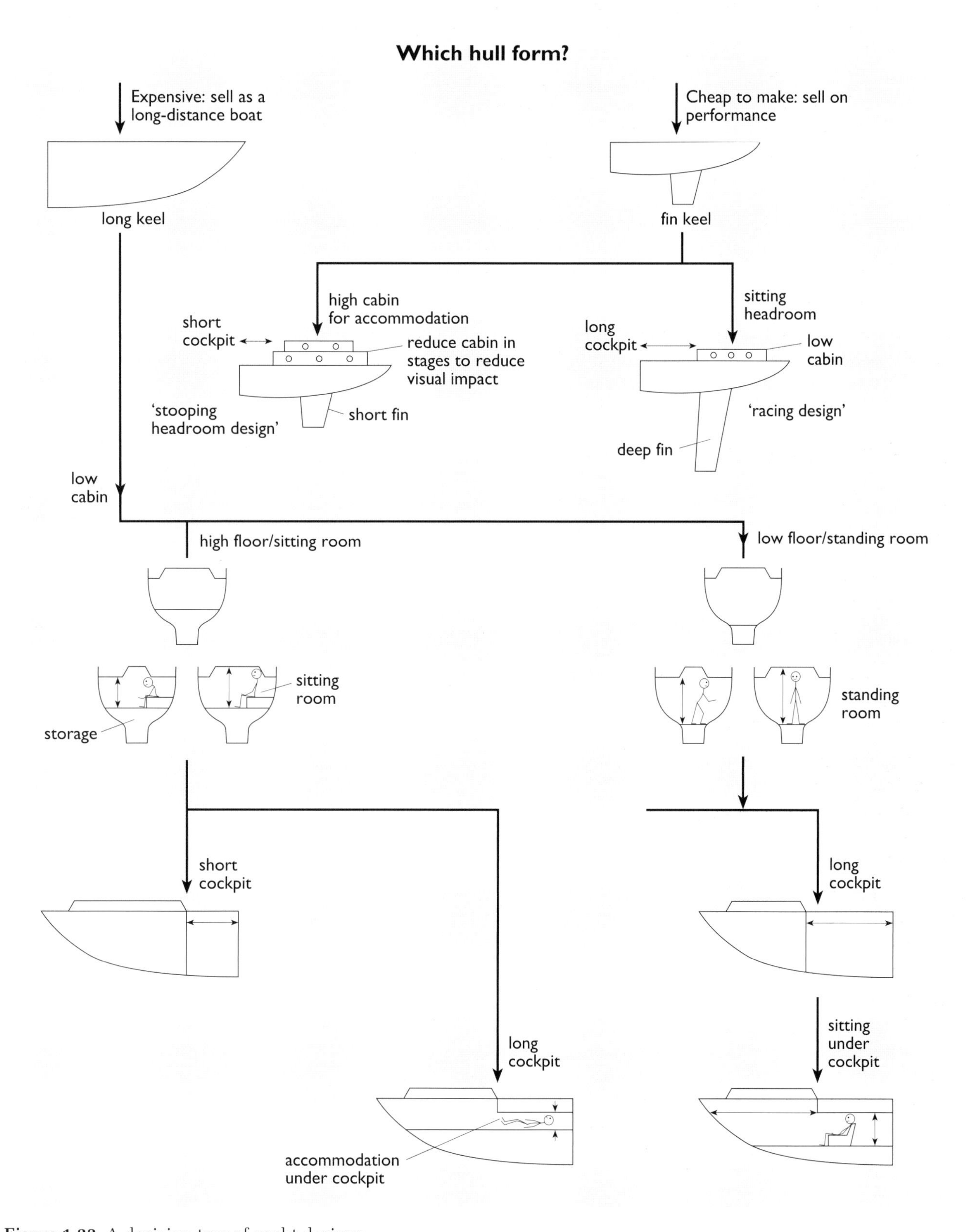

Figure 1.23 A decision tree of yacht designs

that imposes significant constraints on the decisions that follow. The next choice, between a low and a high cabin, on the fin–keel branch of the tree, will lead to very different designs of yacht. The long cockpit, for easy sail handling, naturally goes with a low cabin, for reduced wind resistance, on a racing design. Going for the high-cabin option creates a design where interior accommodation is more important than performance, and a shorter cockpit gives more length in the cabin to split up the space between cooking facilities, lavatories and bunks. It is unlikely that the buyer of a high-cabin boat will want a large cockpit. The low cabin design will give interior space for a sitting person, but not for a standing or a stooping person.

With a long-keel design (top left of Figure 1.23), the extra depth in the hull means that a high superstructure is not required, so the first decision shown is a choice between a high (and therefore wide) floor, giving sitting room, and a low floor, giving standing room. The compromise here is to accept sitting room as a price worth paying for a large floor area that gives more freedom to arrange tables, beds, sinks, lockers and chairs in different ways.

A variable more significant in the high-floor design is the width of the cabin superstructure. A wide cabin gives narrow side decks that inhibit working above deck but might give more headroom inside the yacht. A design where a person can sit underneath a side deck gives a very different use of space than a design where sitting is only possible under the cabin roof.

A further branching of the tree is shown on the basis of a decision about the length of the cockpit and how the room underneath the cockpit floor is used. Shown are two design solutions that show sitting and lying room underneath. Clearly another solution is to have no interior space under the cockpit floor, but to use it all for lockers accessible from the outside. These might be used to store sails and rope.

From this decision tree it can be seen that the design choices about geometry are connected and hierarchical. Also the choices can be made from their implications on the performance of the yacht, without having to model the detail of the aerodynamic and hydrodynamic principles which govern the yacht's performance.

The route taken through this maze of interlinked decisions can lead to yachts of very different types and functionalities.

SAQ 1.10 (Learning outcome 1.7)

How does the initial choice of fin-and-skeg or long-keel design for the hull influence the provision of a superstructure for accommodation?

In the next section, two different design solution spaces are going to be evaluated with the help of some simple formulae.

4.3 Conceptual design for human powered flight: a comparison of two design spaces

It is possible to become fixed in one particular design solution space and not realize the potential of alternative approaches. Figure 1.24 shows Puffin, a human-powered aircraft produced by a design team from Hawker Siddeley at Hatfield that was led by an aerodynamicist. The team produced an aerodynamical solution by subordinating all design decisions to the imperative of a clean airflow over the plane, with no bits sticking out in the wind to cause undesirable drag. The aim of the design team was therefore to extract as much lift as possible from the wings, by having the design as aerodynamically perfect as possible (see ▼Aerodynamics▲). The team even mounted the propeller at the back, so that it would not create any turbulence in the airflow over the wings. The design was made of wood; you can see the detail of the ribs that produce the wing shape and the beam structure inside the wing.

Figure 1.24 The Puffin human-powered aeroplane

Puffin was built to win the Kremer prize for human-powered flight. The requirement for winning the prize was to fly a figure-of-eight course around two pylons half a mile apart, with a minimum height of 10 feet at start and finish, using human power alone. The whole of the historical experience of the aircraft industry was available for the designers to draw upon, and it was believed at the time the prize was first offered – 1959 – that the prize would soon be won.

Puffin flew well enough in a straight line but did not turn well and was a nightmare to repair after crashing.

However, the aerodynamic solution influenced all the subsequent UK attempts to win the Kremer prize. Figure 1.26 shows the Jupiter, a design produced by the RAF in 1975, some ten years or so after Puffin. The concept was the same and the wingspan had grown considerably. Jupiter flew a mile in a straight line, was more difficult to turn than Puffin, and no easier to repair.

▼Aerodynamics▲

An aircraft generates lift – the force which raises it from the ground – from the shape of its wing. A wing shape is shown in Figure 1.25. This shape is known as an aerofoil.

Because of the shape of the wing, air flowing past it, as the aircraft moves forward, has to travel further around the top of the wing than along the bottom. It flows faster along the top edge of the wing. This faster flow reduces the pressure at the top, and the pressure difference between the top and bottom of the wing generates the lift. There is a high pressure at the base of the wing pushing against the lower pressure at the upper surface.

Once the force from the lift is greater than the weight of the aircraft, the aircraft can take off.

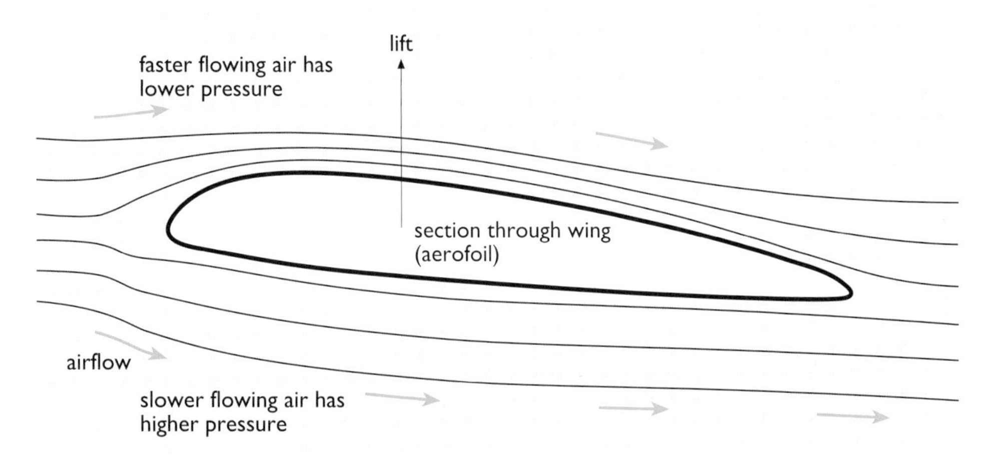

Figure 1.25 An aerofoil

Figure 1.26 Jupiter human-powered aeroplane

So, what were the technical issues and why wasn't the design solution effective?

The continuous power available from a fit cyclist is no more than about 400 W, and losses in the transmission and propeller efficiency reduce this to about 250 W. This is not much – about enough to power a couple of powerful lightbulbs.

Aerodynamics tells us that the power, P, needed to fly a plane is:

$$P = C\frac{M^{3/2}}{\sqrt{S}}$$

where M is the total mass of the plane (and pilot) and S is the cross-sectional area of the wing. C is a number within which is wrapped up the basic aerodynamics relating to flight.

Powers, fractions and square roots are all covered in the Maths Help section of the *Sciences Good Study Guide*.

To keep the power low this equation tells us to build a light plane with a large wing area: two contradictory requirements. A larger wing will mean more weight.

Once the aerodynamic imperative is accepted, that is C is to be kept small, then the structure that is required to keep a large wing strong enough and stiff enough has to fit inside the wing's aerofoil section. The wing spar acts as a beam in bending. The shape can't be varied much, so the main possibility for design choice lies in materials selection. Steel can't be used because it is too dense; wood is a good, cheap, craft choice. As the wing span of designs increased, exotic materials such as carbon fibre were also used. The wing spans became so large that the wings touched the ground under their own weight until the plane was moving and the wing began to generate lift.

The designs that concentrated on aerodynamics achieved the *distance* required to win the prize but none of them turned corners at all well. The problem was that the planes flew very close to the ground because climbing would require more power, and the lift from a wing close to the ground is higher than that at altitude due to the 'ground effect' (an enhancement to the aerodynamic lift caused by flying close to the ground). However, flying close to the ground poses a problem for an aircraft with limited power. Normally a plane banks (tilts to one side) in order to turn efficiently, but these long wings close to the ground could not bank without driving the inner wing tip into the ground, and there was not enough power available from the pilot to climb in order to gain airspace for manoeuvring. Turning with the wing level is aerodynamically inefficient and the inner wing loses a lot of lift because it is moving relatively slowly through the air.

It required an innovation to break out of this aerodynamic solution space which had not achieved the desired result. This was provided by Paul MacCreadey's Gossamer Condor, Figure 1.27.

Figure 1.27 Gossamer Condor human powered aeroplane

We can view this design as occupying a structural, rather than aerodynamic, design space because the design almost ignores the aerodynamic imperative of a low *C*. A structure designed for minimum mass *M* looks totally different. This structural design works more like a boat's mast than a beam, with wires in tension and a rod in compression.

The resulting structure was stiff, strong and light and, at the low flying speeds of the plane, the aerodynamic drag carried insufficient penalty to prevent this design winning the Kremer prize of £50 000 in 1977 at an average flying speed of 13 m.p.h.

There was a good deal of pragmatism in the design. The wingspan was 96 feet 'because the aluminium tube came in 12 foot lengths from the local store' and Condor's mass was 32 kg (70 lb), which was half that of Puffin with its 93 ft wingspan. There were not a great many formers used to keep the aerofoil shape true along the wing, so when the plane crashed it could be patched up quickly.

According to Paul MacCready, the wing spar broke in flight on only two occasions. On one occasion the damage was caused by turbulence from the wake of a nearby crop-spraying aeroplane. Even with half its wing gone, the Condor flew surprisingly well and was able to land gently. Thanks to the wire bracing, the wing did not fall apart at once.

However, the greatest advantage was that the tension wires, which were so structurally important, could be used to twist the wing when the aircraft manoeuvred. The altered angle of attack along the wing evened out the lift when one wingtip was moving through the air much faster than the other wingtip. So, the plane turned level without too much loss of efficiency. This technique was known in the First World War as wing warping, so the Condor

Figure 1.28 Gossamer Albatross. The first human-powered aircraft to cross the English Channel

designer could not claim absolute originality, but it was certainly innovative in the context of human-powered flight.

A further prize of £100 000 was offered by the Man-Powered Flight sub-committee of the Royal Aeronautical Society for the first human-powered flight across the English Channel to France. MacCready returned in 1979 and landed that prize with Gossamer Albatross (Figure 1.28). The aircraft was flown by a racing cyclist, who flew it across the 38 km in 169 minutes, despite an unexpected headwind.

Exercise 1.5 (Revision)

What was the average speed of Albatross in

(a) kilometres per hour,

(b) metres per second?

The second Gossamer design was a logical development of the first, with some different materials and a few advertising posters.

Why was Condor so successful? Intuitively, one would think that the best solution would be the one which is aerodynamically best, rather than the one which is most structurally sound.

Using the equation I gave earlier, we can perform a rough calculation to compare the designs of Puffin and Condor.

What we know is:

Puffin had a mass of 63.5 kg and a wing area of 36 m^2.

Condor had a mass of 34 kg, a wing area of 75 m^2 and a pilot mass of 61 kg.

First we notice the success of the structural design solution. Condor has twice the wing area for half the mass. Admittedly the Condor wing is less efficient, but that may be a price worth paying. Let's assume that the same pilot flies both machines (interestingly the aerodynamic designs such as the Puffin were flown by experienced pilots, whereas MacCready chose a racing cyclist). Thus the pilot will be assumed to have a mass of 61 kg in both cases.

Exercise 1.6

For each of Puffin and Condor, use the formula on p. 38 to calculate the power required for them to fly. You will not know C in either case, so just produce a formula of the form, for example

$$P_{\text{Puffin}} = C_{\text{Puffin}} \times x$$

where x is the value of $M^{3/2}/\sqrt{S}$ for, in this case, Puffin.

If the two planes are aerodynamic equals, then C will be the same for both. They are not, but aerodynamic improvements at low flying speeds tend to be small percentages so there will not be an enormous difference. The answer from Exercise 1.6 shows that, if C is similar for each plane, Puffin needs twice the power to fly than Condor. Improvements to the aerodynamics could not generate the extra lift needed to overcome this barrier.

4.4 Conclusion: the importance of concept

In this section we have looked at how the generation of an initial concept can dominate the design decisions which are made thereafter. The amount of space available in a boat is dependent on the keel shape; the decision to try for aerodynamic perfection in an aircraft may increase the power required to fly it.

The process of generating concepts is important for two reasons.

First, it helps to define the *design space* (that is, the 'problem space' and the 'solution space') which may be returned to and re-examined via various iterations in the process. All details are not defined; there is still room for variation, decision and choice. This freedom identifies the design space. Of course this freedom is limited and concept design has defined the limits of this freedom at subsequent stages in design.

Second, in any practical situation it is ultimately very important to get the concept 'right'. Remember we are limited at future stages by the concept chosen. This is sometimes encapsulated in a piece of design 'wisdom' that says 70–80% of product costs are determined at the concept stage. This is not to say that generating concepts costs a lot in time and money, but that as the design progresses to manufacture and market there is little that can be done to influence the costs incurred. Hence the importance of reducing weak ideas early in the process.

SAQ 1.11 (Learning outcome 1.4)

(a) In Section 1 I discussed various types of model, such as scale drawings, three dimensional models, and so on. How do such models assist in the definition of the 'design space'?

(b) What type of models might be used in the conceptual design of sailing boat hulls? (Think in terms both of the form of the hull and the overall performance of the boat.)

In the following sections, I'm going to look at examples of how an initial concept can be carried through to manufacture, via a prototype.

5 Concept to prototype

5.1 A process of focusing

In the last section I examined one part of the design process in detail: the generation of concept. Towards the end of this stage, ideas for a design are given some detail. However, they are still not far enough advanced to be made as a prototype, nor can they undergo rigorous testing. They are still *ideas*. They will most likely take the form of two-dimensional models such as sketches, renderings or outline plans but they may also be manifest in three-dimensional rough models: usually scale models, but perhaps also full-size models where feasible. In some situations the concept designs will be available as digital models in a computer.

Many features, although not developed in detail, should work in principle. This 'in principle' is a little misleading since it contains several aspects. One aspect is theory, which indicates by rough calculations, possibilities, sizes, shapes and materials. Another aspect is experience; something that has worked well before in similar circumstances should work again.

As we saw in the last section, the process of generating concept is important. It helps to define the design space. Once we get the problem space and the solution space accurately mapped we are less likely to come up with a weak or unworkable design.

However, we should not be carried away with the importance of concept. Designs are ultimately used. The concepts have to be turned into useful things. The generation of concept has helped to focus on a particular type of design; perhaps the hull shape for a boat. We now need to look at the ways that the details are completed so that we can define exactly what our design is. It can then be manufactured and tested.

5.2 Down to the detail

A little later I will describe an example of a product undergoing this process of defining the detail where you can follow the decisions that are made. First I will give some general examples to give a feel for the range of design activities used in transforming concept into prototype.

First think of a jet engine used in a commercial aeroplane. There are three major designers and manufacturers of these in the world: Rolls-Royce, General Electric, and Pratt and Whitney.
All three are trying to apply new science and engineering to making engines that are more powerful, more efficient and quieter. Figures 1.29 and 1.30 show a jet engine to give you some idea of the scale and complexity of this product.

A design team will have made decisions on materials, configuration of inlets, blades, burners and exhaust, temperature of operation, size, and power output.

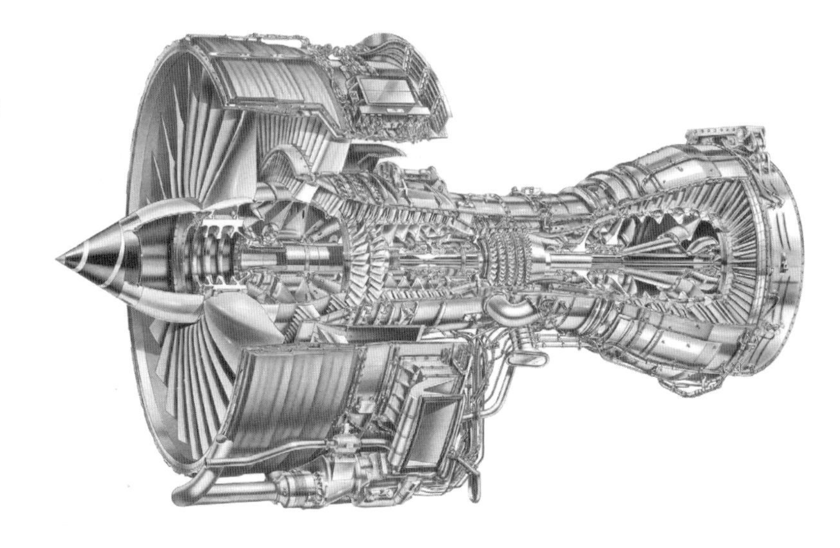

Figure 1.29 Cut-away of a Rolls-Royce Trent 600 engine

Figure 1.30 Rolls-Royce Trent engine

Some of these will be determined by specific needs in the industry tied to the operating conditions, types of aircraft, payload and range. At this stage there are many ideas incorporated into a concept. For example, a particular material may be chosen but there may still be uncertainties as to how it is to be manufactured in the shapes and surface finishes required. To run the jet at a high combustion temperature may require cooling channels in the blades. Maybe this has been done successfully before but perhaps not at the temperature needed for a new jet. The loads on bearings for all flight conditions may not be known accurately, but based on experience and preliminary calculations the design team will have defined a range of sizes and shapes.

There still is considerable design work to complete in defining (i) shapes and surfaces, (ii) geometry and configuration of cooling channels so that they can be produced without weakening the blades and (iii) specifying bearings (probably for detail design and manufacture by a third party supplier). There are many possibilities for each of these which need exploring. Models which use mathematics, computers and practical testing are used to determine the performance of different possible details for the designs.

Think of a new car. A concept design for a new model will cover areas such as shape and initial styling (see Figures 1.4 and 1.6), and engine configuration, suspension and transmission. The types of these may be suggested, but not in any detail. There is still major design effort needed to transform a body style into a pressed and welded bodyshell; to specify the gears, bearings and shafts in the gearbox for the operational loads; to specify engine components for performance, efficiency and cost.

Exercise 1.7

Suggest three reasons why defining the details of a design can take a long time and considerable effort.

It is at this stage between concept and prototype that models of designs are widely used. Designers need to predict the performance of possible designs without building a complete prototype at large cost: a prototype of a vacuum cleaner may be economically feasible, but a prototype of a jet engine is less so. The predictions will not always be entirely accurate because as we have seen, models are always incomplete in some respects. The models do not include all the factors which influence the design when in use. However, they are a help and guide in progressing through this stage of design. The models help designers to make choices among possibilities in the design space.

Sometimes this part of the process of design is seen as rather routine. The excitements of looking at new concepts or the building and of testing of prototypes are real and tangible. However, there is substantial scope for creativity. Just because the concept stage has defined limits on possible designs does not mean that this part of design is boring and mechanical. Making good decisions in developing detail depends on understanding the close relationships among different parts and features of a design. Balance, compromise and judgement are needed. This is equally creative as the grand schemes and broad brush at the concept stage.

So in trying to define the details of a design we might try many possibilities. Some parts of the design may be right and others wrong. As a result of putting right one part, another part may go wrong. Designers try and learn from this exploration of design possibilities and 'home in' on a good design.

In order to look at design possibilities at the concept-to-prototype stage of design, you might like to look at a game, ▲The detail design game▼. This is not real design of product (we will examine a real product later) but it is an illustration of some of the ways of thinking that designers use.

5.3 Design and innovation 2: the 'Res-Q-Rail' stretcher

One of the difficulties with designing is that it is almost always complex. There are contexts, which impose particular constraints on parts of the design and their connections. To describe a design case study, especially for the core stage between concept and prototype, it is necessary to be quite clear about the context in which the design takes place, the stage of definition the design has reached, and finally any requirements of the users and clients which are relevant.

I will describe here a design for a device which is used to transport equipment and casualties to and from the site of railway accidents. A particular difficulty with railways is that they pass through remote areas, with no alternative means of overland access. Fire and ambulance emergency services when called to the scene of an accident may only be able to gain access to the track a mile or so away. These emergency services require a lightweight, portable and compact trolley to take heavy breathing apparatus and cutting equipment to the scene. Casualties on stretchers then need to be transported to waiting ambulances.

This design started life as a requirement from the emergency services in Northumberland where the railway line to Scotland is inaccessible for long distances. Figure 1.37 shows an emergency rescue team carrying a stretcher alongside a track in a simulated training accident. Note that six people are needed and progress is hazardous.

Figure 1.37 Carrying a stretcher beside a railway track in a simulated training accident. Four people are needed to carry one stretcher plus two to attend to the casualty

▲The detail design game▼

Suppose a design is to be made which consists of three modules, and within each module there are several components. Each component is a simple rectangular block. Within modules, the components are connected end-to-side as shown in Figure 1.31.

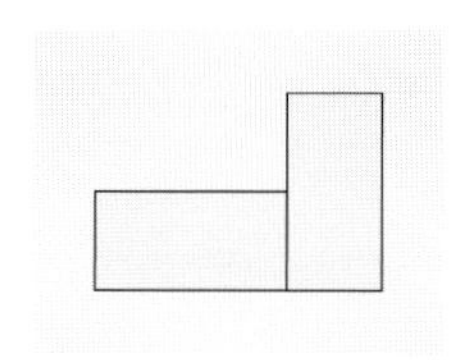

Figure 1.31 The allowable connection: end of one component connects to the side of another

Connections between components within modules can *only* be end-to-side. There should be no accidental edge-to-edge connections between components in any other way. For example, the connections between components shown in Figure 1.32 are not allowed inside a module. We must make sure that these 'invalid' connections do not accidentally arise whilst making valid connections.

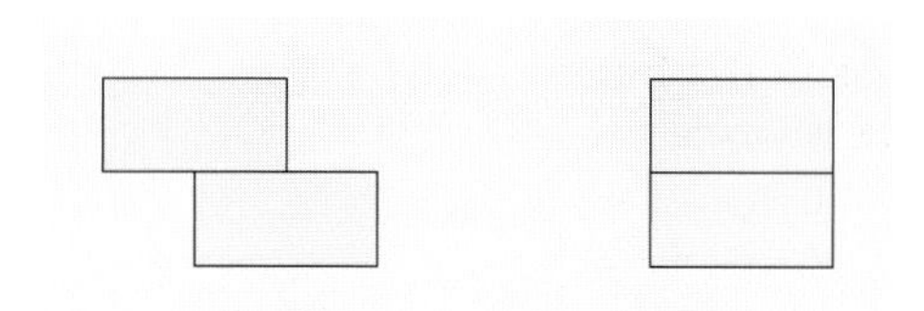

Figure 1.32 Invalid connections between components

Connections *between* modules, however, are made by connecting components inside the modules side-to-side, as in Figure 1.33. There must be no accidental connections between modules in any other way.

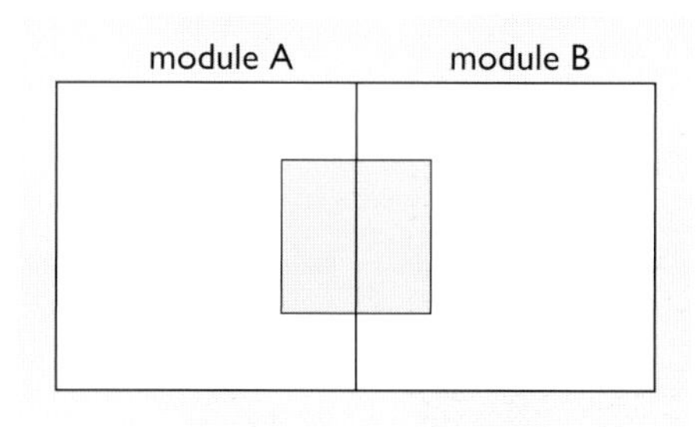

Figure 1.33 Connecting components between modules

Suppose that a conceptual design has determined that three modules are needed to be connected in a row. There must be one component in the top left corner and one in the bottom right (this might be for connecting this group of modules to another group). Figure 1.34 shows this arrangement.

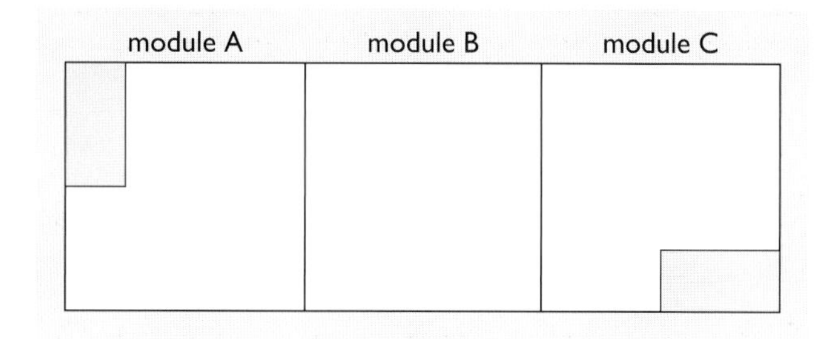

Figure 1.34 Starting point from the conceptual design

We might start in module A as shown in Figure 1.35. This has three components inside the module. Note that each module does not need to be entirely filled with components. (You might ask at this stage whether there are any other ways to connect components in Module A.)

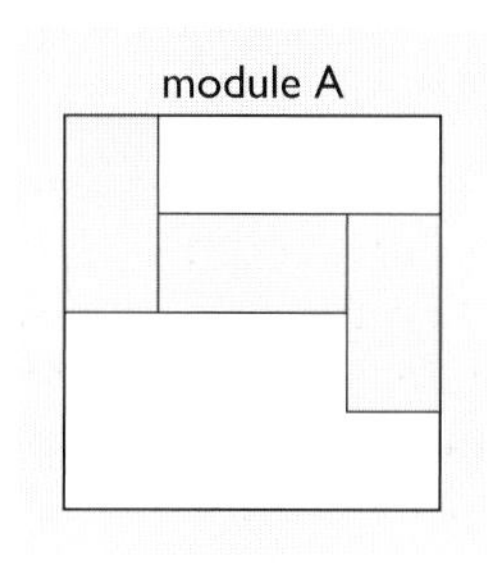

Figure 1.35 Three components in Module A

To connect Module B you need to add the first component in Module B along the long edge of the third component in Module A, as in Figure 1.36.

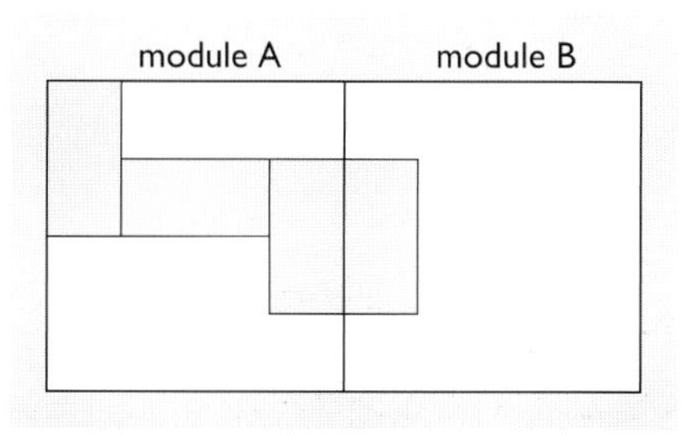

Figure 1.36 Adding Module B

The design problem is to connect components in each module and connect modules according to the allowable connections. There are potentially many solutions.

SAQ 1.12 (Learning outcome 1.8)

(a) Try to find one solution to the game by sketching layouts of components on Figure 1.34. The aim is not to find the best solution but to reflect on how you obtained a solution.

(b) Reflect on how you arrived at your solution. How did you go about solving the problem?

The concept of the 'stretcher carrier', as it was called initially, had five modules. These are shown in Figure 1.38. They consist of:

two cross pieces (front and back), with wheels, for stretchers to rest on;

two detachable side pieces joining the front and back;

a canvas sling between the two sides.

This modular construction allowed the stretcher carrier to be stored compactly, transported to the site easily and assembled quickly.

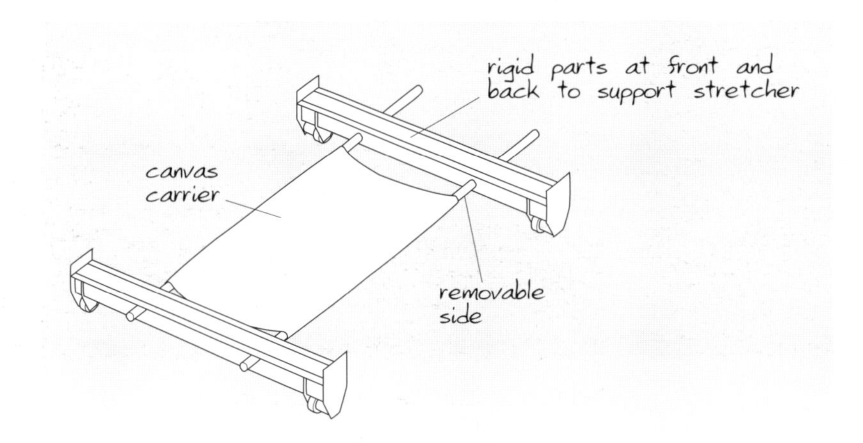

Figure 1.38 Stretcher carrier – concept

This concept was developed to a prototype in 1994 by designer Rob Davidson. It is patented and in production. The annotated drawing in Figure 1.39 shows the detailed features in the final design. The annotations explain the reasons for the features introduced when developing the concept.

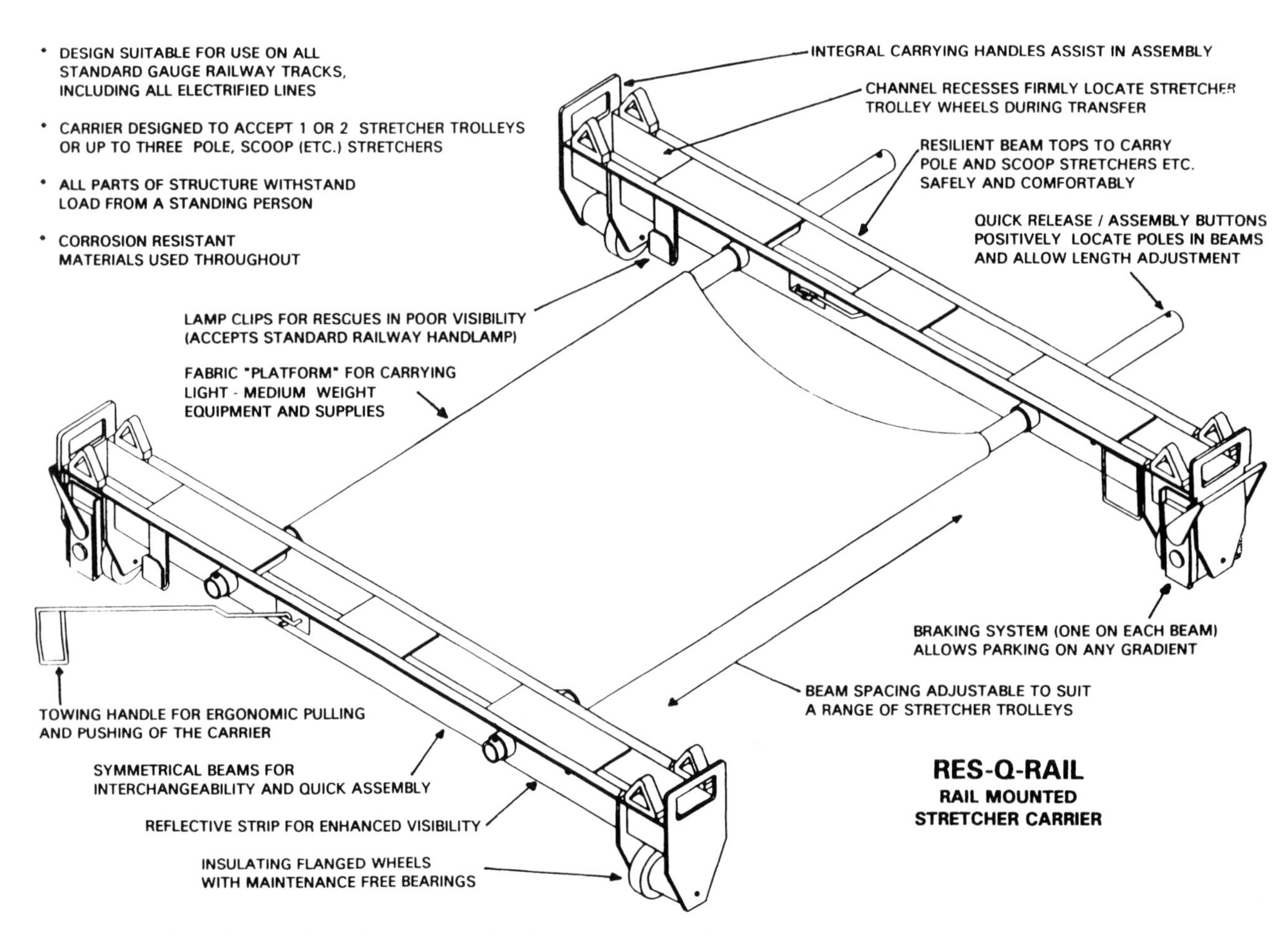

Figure 1.39 Stretcher carrier – the concept developed

A prototype was built according to the developed concept and tested. Figures 1.40 and 1.41 show tests of the final production stretcher carrier now called 'Res-Q-Rail'. Figure 1.40 shows tests carrying stretchers and Figure 1.41 shows tests carrying equipment and people.

Figure 1.40 The Res-Q-Rail in operation

Figure 1.41 The Res-Q-Rail in operation

I wish to draw your attention to two things. First, the design of the front and back modules uses an aluminium beam fabricated by bending thick sheet. This is strong and lightweight. Second, the wheels are plastic for light weight and low friction. However, nylon, often a good choice where lightness and low friction are needed, is not used. Nylon expands when wet, which might cause the wheels to jam.

The first point – the use of folded aluminium sheet – represents a general design principle of making strong things cheaply, from simple materials. The second point – concerning the material for the wheels – illustrates a specific piece of technical knowledge. There are, of course, a multitude of such guides and constraints on getting to the prototype in this example.

Aluminium was the preferred material as it is both corrosion resistant and lightweight. However, for the carrier, rather than a rigid, flat surface, a sling design is used. This is light and flexible for packing, has a natural dip to hold equipment, and is easily assembled by inserting the side tubes.

After a prototype had been developed, the design was further refined and put into production. However, as with many designs this was not the end of product development. In use, fortunately in training rather than at accidents, a number of new requirements emerged. One was that it was important to keep the carrier stationary, and thus a brake was requested by customers.

Brake concepts were examined by the designer, who decided on a lever design operating by friction onto the track. The key requirement was to transform the vertical motion of a lever (like a car handbrake) to the sides of the carrier and then onto the track. Remember that the wheels are plastic for lightness and corrosion resistance so that it was not feasible to brake onto the wheels as plastics may be easily deformed.

The overall concept of the brake is outlined in Figure 1.42. The up/down rotation of the brake handle is transferred through a mechanism M to rods L_1 and L_2. These rods act to pull the brake pads P off the track by means of another mechanism N. The 'natural' position of the brake handle is on. Positive action (pulling the brake handle up against a spring) is required to pull the brake pads away from contact with the rails. Thus the brake is conceived quite differently from a car handbrake which requires effort to apply the brake. In this case the effort is needed to take off the brake.

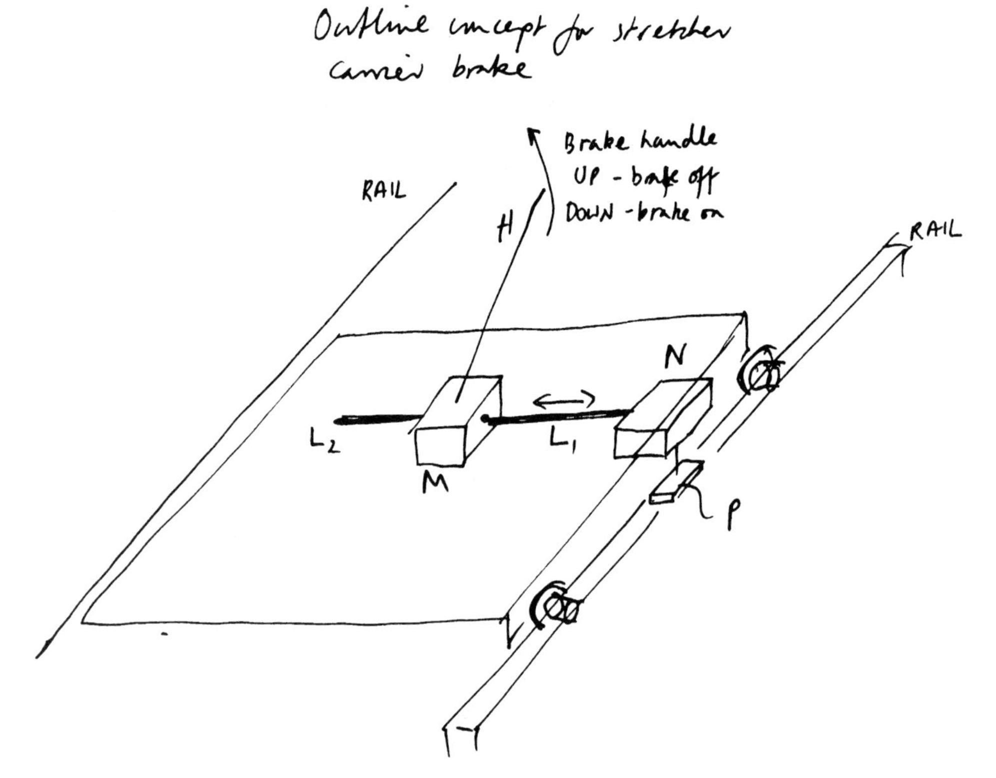

Figure 1.42 Outline concept for stretcher carrier brake. H is the handle. Pulling up H raises brake pad P against a spring. Releasing H causes P to spring down against rail. Mechanism M converts up/down motion of H to side-to-side movement of L_1 and L_2 (L_1 and L_2 are link rods). Mechanism N converts motion of links L_1 and L_2 into up/down motion of pad P

We will look at one part of this design, namely the mechanism M for converting up/down motion of the brake handle into side-to-side movements of links L_1 and L_2. The spring loading of the brake will be incorporated in the mechanism N. We will only be concerned with the brake handle mechanism M.

The question now is how to realize the concept shown in Figure 1.42, particularly the mechanism M. Figure 1.43 shows a sequence of developments for the mechanism M using sketches made by the designer whilst in the process of designing. There are several possible developments, including gears, that were considered by the designer, but we will concentrate on just one of these. This was the preferred route adopted by the designer. Six stages are shown down the left of Figure 1.43 as a sequence of sketches and drawings made during the design. The first five stages are rough sketches made by the designer and the last stage is a computer aided design model of the detailed design. On the right of Figure 1.43 are two further sketches relating to the design sequence on the left. Note that these are not enlargements, but further refinements of the design.

Notes are included in each stage indicating how the designer was thinking. Do not be concerned if you cannot read them. These are notes transcribed directly from the designer to show the design 'in progress'. As these are the designer's sketches and notes there are many things that are not being explained. I use this series of sketches to reveal something about the process used rather than for you to have to achieve a detailed understanding of how the brake works. There are several engineering 'judgements' made about tolerances and deformations which are needed to make a design like this work. The designer will only discover whether these are right when prototypes are tested.

Figure 1.43(a) shows the brake handle in two possible locations – either pointing forwards or backwards. We may want someone to operate the brake

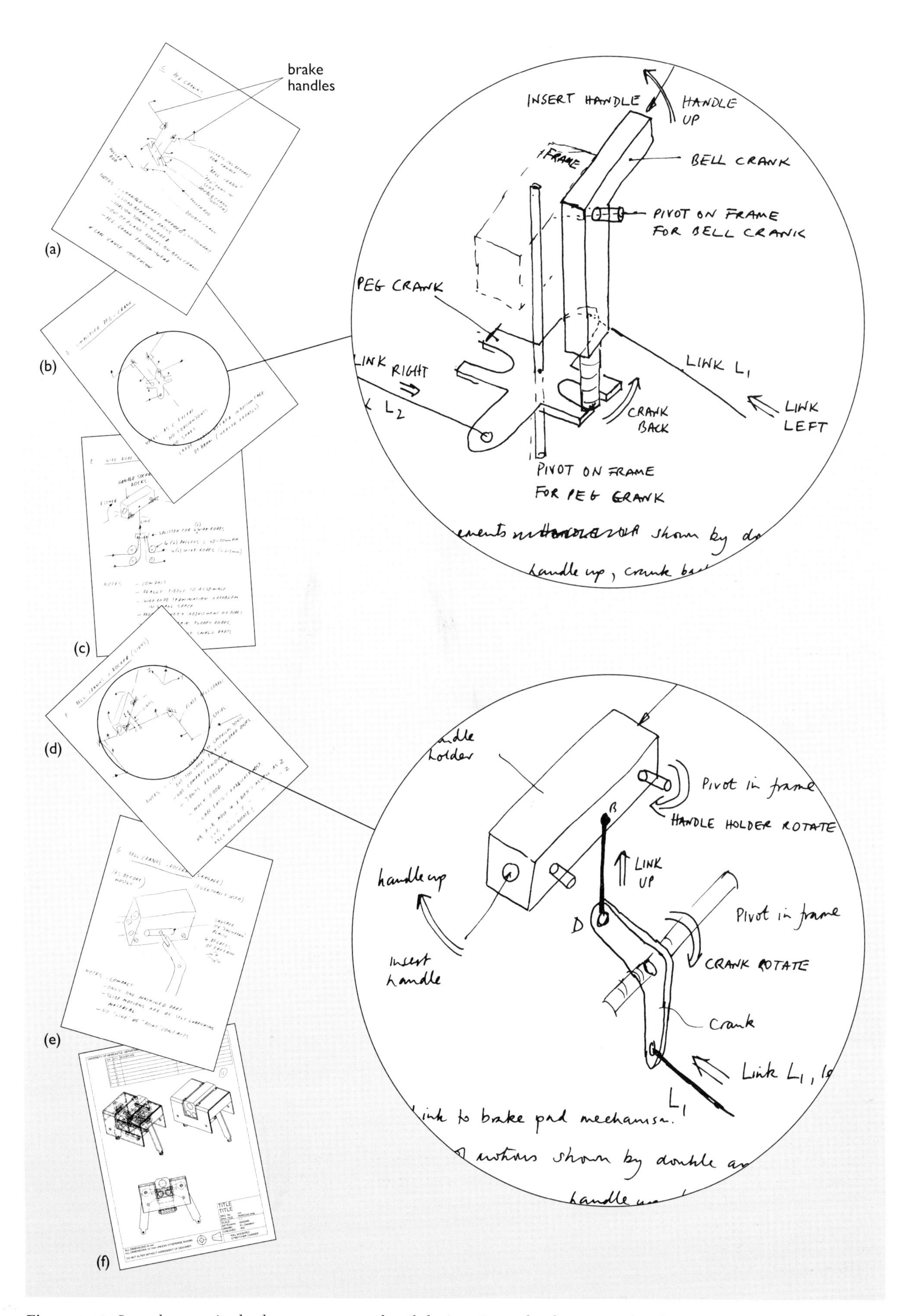

Figure 1.43 Stretcher carrier brake – concept to detail design. From the designer's sketches

from on the carrier (when transferring equipment and people to the accident scene) or from a position off the carrier (when transporting casualties away from the scene).

The handle is detachable so that it can be inserted into mechanism M in either the forward or the backward position. The sketch shows a mechanism with two handles, that is with one in the forward and one in the backward position. Remember that only one handle is used at a time.

Look at Figure 1.43(b), the second sketch in the sequence with its companion drawing and annotations. A 'bell crank' is attached to the handle. This crank is pivoted on a fixed frame which will be attached firmly to the front or back cross piece. The fixed frame will probably be in a block containing the mechanism. This will be attached to the cross members of the stretcher carrier which run between the wheels. The end of the bell crank then moves backwards or forwards in slots in a rotating 'peg crank'. The rotation then pulls the rods which operate the brake on the track. We are not going to deal with how the brake is brought down on the track. We have quite enough complexity just getting this bit of the design sorted out. Note that the next sketch, Figure 1.43(c), shows a cable operated solution which was not developed further.

The sketch in Figure 1.43(d) is getting close to the final design. All the essential elements are present. Look at the associated drawing. There is a frame for the mechanism which is to be fixed to the stretcher carrier, probably on one of the cross members. There are also two pivots on this frame, one for the brake handle and one for the crank. Compare the two designs in Figures 1.43(b) and Figure 1.43(d). They are very different.

The final design is not a sketch. It is an accurate computer-generated picture. At the time of writing (autumn, 2000) this is a current 'live' design and a prototype has not been built or tested. Many details still remain, such as a ratchet to keep the brake handle up, the mechanism for raising the pad and spring loading the brake.

The sequence of possible developments of concept in Figure 1.43 follow a kind of design logic. Changes and transformations of the design do not occur randomly. At each stage the designer considers the current design, looks for opportunities to develop it, perhaps by changing a shape, combining separate components into one, or adding a new feature to meet user requirements. However, the designer is constrained. The design must be easily manufactured and must be very robust (the design will be treated roughly by people in a hurry in difficult conditions).

This is a typical design problem in the engineering industry. There is no space science here (you will see some of that later in the course). This is an example of designers being creative in meeting a need. It is not a mundane problem. It needs creativity, thought, logic and intelligence to bring such a product to market. We have examined an example which at the time of writing has not been tested. This is intentional, in order for you to catch a glimpse of the middle of the design process without the benefit of hindsight which tends to sanitize the design process as a smooth and 'logical' progression from need to concept and prototype. In reality the process is messy and creative as people try to come to terms with a problem and possibilities for its solution. Designing is a complex process, there are guesses, blind alleys, failures and successes. This characterizes all stages of design.

SAQ 1.13 (Learning outcome 1.9)

There are many design failures in the things we use every day. We notice these more than the easy-to-use, naturally pleasurable products of good design.

(a) Identify a simple product or part of a product you think is poorly designed.

(b) Describe the poor design and give an initial suggestion or concept for improvement. This is not an easy question. Just try and think through the steps. I want you to realize how difficult these questions are and recognize that designing is hard! Do not worry if you cannot complete the answer.

What SAQ 1.13 should indicate to you is that problems and poor design do not always stand out clearly. We all get accustomed to using designs which are far from ideal. We all adapt. We have to think about possible solutions to realize the extent of design shortcomings. In a curious sense design-minded people are evaluating all the time; they are continually exploring what might be. Good designers see new (and possibly better) ways; poor designers put up with old ways. So you could say that all design is innovation. I will come back to this theme towards the end of this part of the block.

The next stage of design takes a prototype, probably a rather rushed and incomplete affair, and submits it to testing so that its performance can be evaluated. As a result modifications are made and the final design delivered to market. This is the subject of the next section, which describes a case study of taking a design to market.

6 Design and innovation 3: the Brompton folding bicycle

6.1 Reprise: concept to prototype to production

Go to buy any functional product, and you will almost certainly be presented with a range of different designs. Some of the differences will just be in the styling, but there may also be real differences in function or quality, which may be reflected in the price. Different design concepts lead to competing products with particular sets of advantages and disadvantages. Moving from concept to production depends critically on the industrial and social context. An idea for a new product, or a modification to an existing design, requires both human effort and financial input if it is to come to fruition.

Part of the design process is the development of prototypes. A prototype is a 'test' version of the product, and may have different functions depending on when it is constructed during the design cycle. If the product is simply having a change to its styling, the prototype will be important in establishing the 'look' which will be attractive to consumers. If a new piece of technology is being used to improve a product, the job of the prototype may be more technical: to ensure that the product's performance is up to scratch. Prototype development may be one of the most costly and time-consuming stages of finalizing the design; it may involve extensive market research, or prolonged laboratory and consumer testing.

If the design life cycle is shortened, to hasten the arrival of the new product in the marketplace, the risk of failure goes up. More designs for a product arriving faster on the shelves is good for consumers, who will revel in the choice, but not good for employers or employees who are staking money and jobs on success!

As an example, James Dyson is on the record as saying that the design of his cyclone vacuum cleaner came about after the making of 5000 prototypes.

The third case study I have chosen to continue the design story is an accessible example that allows me to look at some engineering specifics: it is the design and successful production of a folding bicycle. At the end of the study I shall consider the general lessons and issues that arise from the study. However, remember that most designs fall by the wayside, so its success makes it atypical.

6.2 Bicycle origami

Andrew Ritchie started designing a folding bicycle in 1975, stimulated by the Bickerton folding bicycle design. The Bickerton (Figure 1.44) is made from aluminium, and is hinged at the chainwheel bracket. (The chainwheel is the toothed wheel driven by the pedals.) This means that the chain and chainwheel are on the outside when the bicycle is folded, and the two wheels come together.

In essence, Ritchie was inspired by the thought that he could do better. His two major criticisms were that the bicycle didn't fold well because the chainwheel, the muckiest part of a bicycle, was prominent; and that he did not think that aluminium was the best material for a folding bike:

> Aluminium is too soft for a folding bicycle, it just doesn't stand up to the knocks, the everyday wear and tear.
>
> Ritchie (1999)

Figure 1.44 The Bickerton, a source of inspiration

The first criticism is easy to accept, but his view on aluminium is not at all obvious. After all, many bicycles are made from aluminium, which is a light, corrosion-resistant material, seemingly ideal for a portable bicycle. If it is good enough for top-of-the-range cars like the Aston Martin V12 Vanquish, why not for a bicycle also? Remember that its corrosion resistance and low weight made it a good choice for the Res-Q-Rail. I shall return to this issue later.

In an existing firm, say a bicycle company, an idea for a new product such as this would include other people in critical roles. Perhaps market researchers, involving bicycle users, to estimate the size of the potential market and the interaction between designers with technical expertise in, say, production and structures. Cost would play a large part in the discussions, as would risk and the effect of the project on existing products and commitments.

An independent designer can often find it difficult to get a sympathetic hearing when they take their ideas to established manufacturers. They face the 'not invented here' syndrome, which suggests that companies put their faith in their own in-house ideas but cannot see the potential in ideas from outside. Alternatively, they see potential legal and economic problems in protecting and investing in a design which may have been shown to competitors. This is a common enough story: the Dyson vacuum cleaner was hawked around established vacuum cleaner companies who rejected the idea. Andrew Ritchie was to experience the same rejection from bicycle manufacturers.

Figure 1.45 Designer Andrew Ritchie with a folded Brompton bicycle

His basic idea, which remained constant through the development of prototypes, was to hinge the bicycle to make the wheels come to the 'centre', one on each side of the chainwheel. In this way the wheels would shroud the oily chain and chainwheel.

Such a 'kinematic' solution (referring to the way that the parts of the bicycle move relative to each other) occupies a different design space from that of the Bickerton. It gives the same functional solution – reducing the length of the bike down to something which is more portable – but the way by which this is achieved is different. The concept of where the bicycle is hinged, and how its parts are arranged when folded is different. Once that concept is established a way of realizing it is required.

By way of introduction to the Brompton story, Figure 1.46 shows a recent production Bromption being folded. The first stage is to swing the rear wheel underneath the frame – see pictures (a) to (d). As you can see, the wheel hinges in its own plane.

The next stage is to move the front wheel to a position alongside the rear wheel; pictures (e) and (f). This is done by freeing a clamp on the frame crossbar, near where the crossbar joins the headstock. (The headstock is the part of the frame to which the handle-bar pillar and the front wheel are attached.) Once the clamp is freed, the front of the frame can be hinged sideways to bring the front wheel beside the rear wheel. This sideways movement of the front end is a significant feature of the production model. As we shall see, the first prototype Brompton used a different technique.

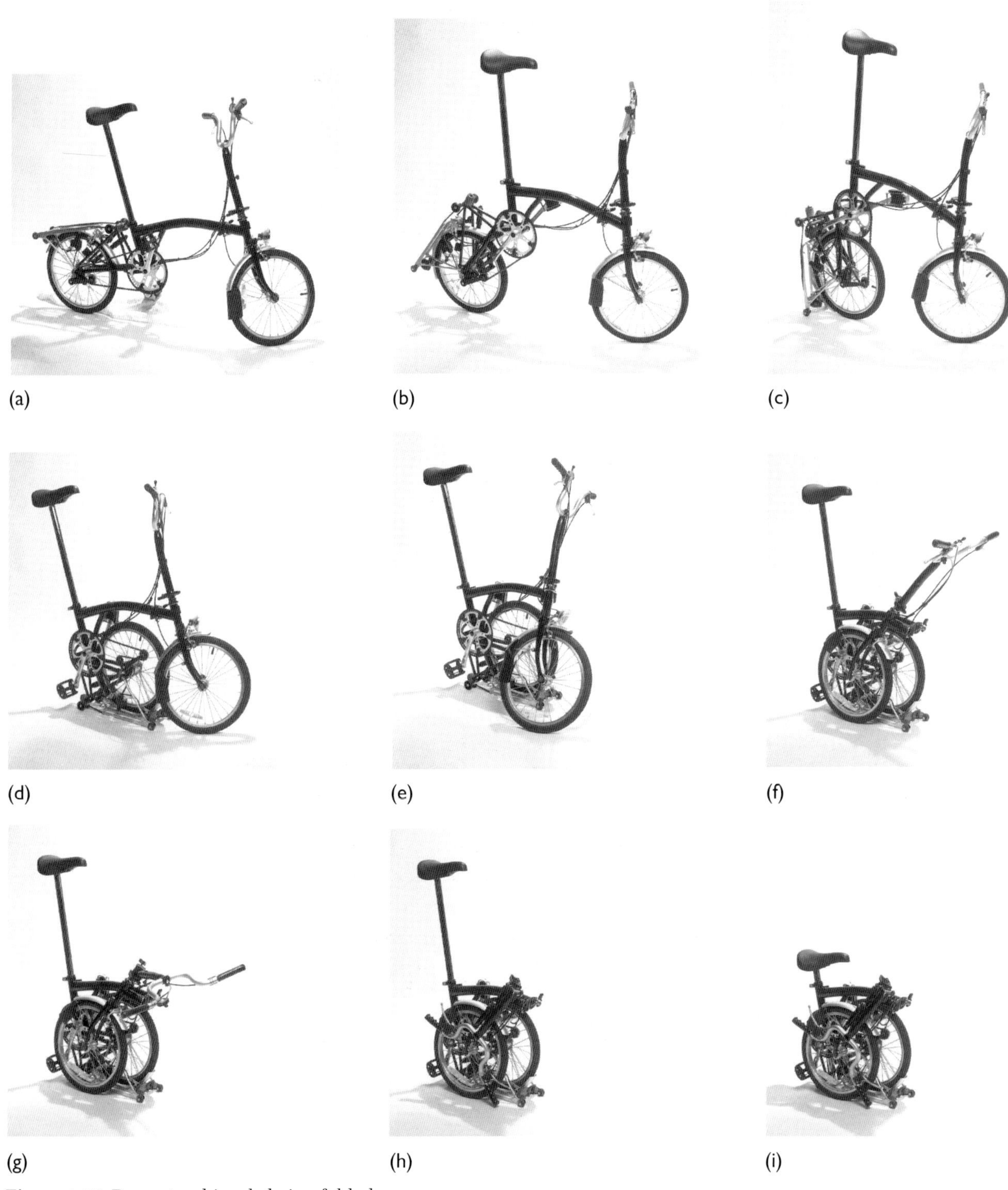

Figure 1.46 Brompton bicycle being folded

(a) (b)

Figure 1.47 Hinges and clamps

Another clamp, just above the headstock (Figure 1.47), is freed, allowing the handle-bar pillar to hinge down to sit alongside the front wheel; pictures (g) and (h).

The final step is to unclamp the seat pillar and to slide the seat down; picture (i). This action locks the bicycle into its folded arrangement.

In Figure 1.47(a), the clamped crossbar hinge is visible behind the bundle of cables. This hinge, when unclamped, allows the front wheel to be moved to a position alongside the rear wheel, as in Figure 1.46(e) and (f).

Bicycles designed to be folded into a convenient shape have a long and honourable history going back at least as far as 1885. Figure 1.48 is a collage of a few of the many solutions to the problem. Common to all these designs (and the Bickerton in Figure 1.44) is the problem of the protruding chainwheel, so Ritchie's concept looks to be a genuine innovation.

6.3 Prototyping and improving

In Ritchie's first prototype design (P1) the rear wheel hinged forward in its own plane from the lowest point of its triangular support structure, as in the production model in Figure 1.46. However, unlike the production model, the front wheel of P1 also moved (almost) in its own plane underneath the bicycle to sit alongside its partner; in this case some sideways movement was needed to ensure that the front wheel sat next to the rear one, rather than just bumping against it as it hinged. To achieve this the front wheel needed a complex, skewed hinge to move it the few inches sideways so as to clear the rear wheel and chainwheel.

As well as moving the two wheels to the centre, it was necessary to move the saddle, together with its pillar, and the handlebars into the same space. The seat pillar telescoped to get the saddle into the packing space, which had the advantage that saddle height adjustment and packing were accomplished by the same mechanism. The telescoped seat pillar slid down behind the hinged rear wheel, so locking it in place, an important feature that has survived the transition from prototype to production.

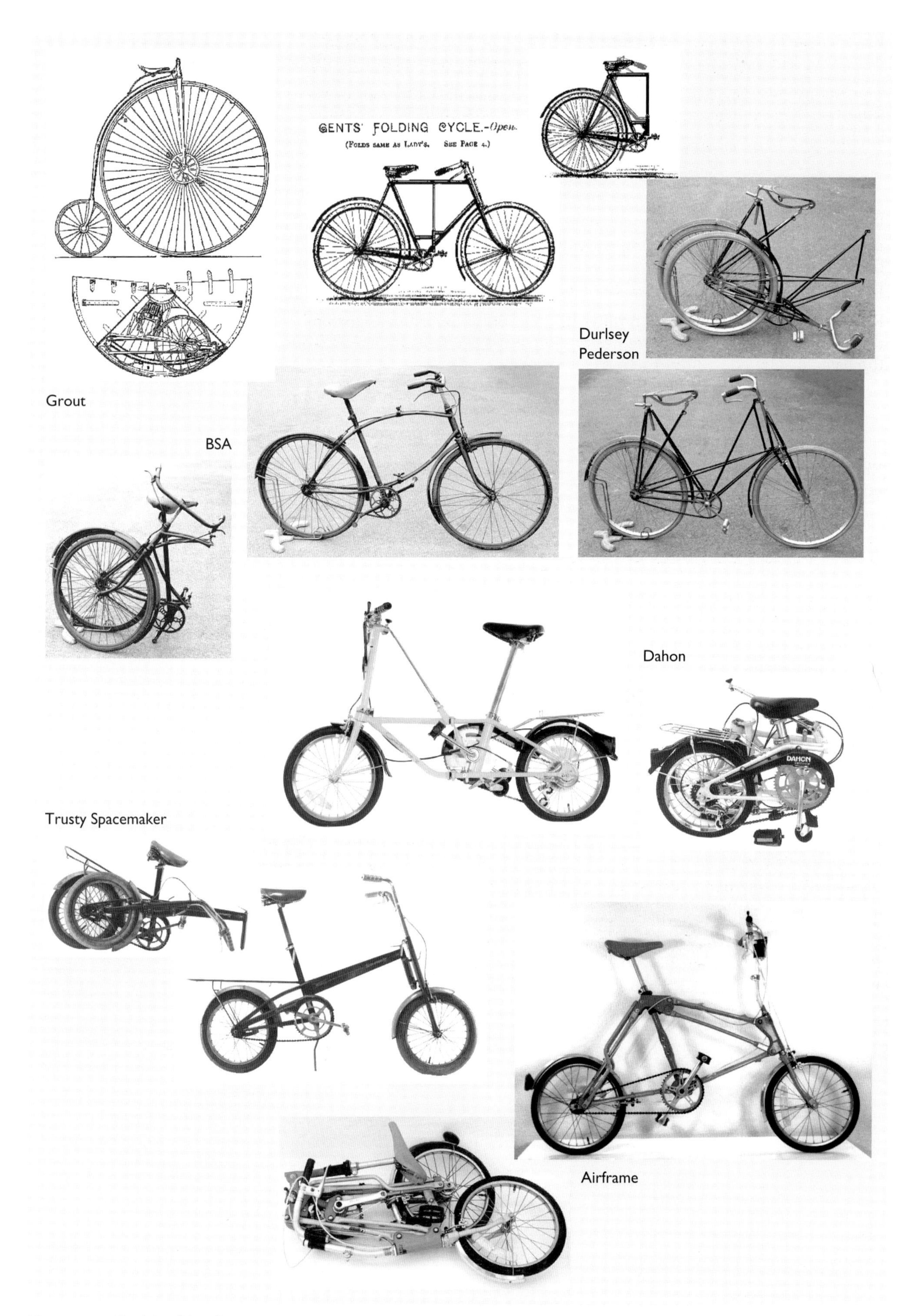

Figure 1.48 Packing bicycles

Ritchie was driven by a search for 'the ultimate in compactness' when designing and building P1, which was a platform for various design ideas.

The chainwheel and the saddle competed for space in the folded package, so Ritchie tried to move the chainwheel away from where the saddle needed to be, but

> ...it was too complicated, I gave up an inch when that idea was dropped.
>
> Ritchie (1999)

Prototype P1 used 18 inch wheels, then common on children's bicycles. The main tube of the frame was lower than in the production model and the bicycle was not stiff enough (see ▼Stiffness and flexing▲). Bowden cables linked the front- and rear-wheel folding mechanisms.

Ritchie is a regular bicycle commuter in London, so he tests designs and design changes routinely and expertly. He was pleased with the realization of the basic design concept in the first prototype:

> I had demonstrated that the design concept could result in a compact folding bicycle.'
>
> Ritchie (1999)

Ritchie uses the expression 'good luck rather than design' to describe unpredicted advantages of his conceptual design solution.

6.4 The second prototype (P2)

The major design difference between P1 and subsequent prototypes was the removal of the complex skewed hinge required to move the front wheel in its own plane underneath the bicycle to sit alongside the rear wheel. The front wheel now hinged orthogonal to the plane of the bicycle (i.e., it moved sideways from the line of the frame, as happens in the production model in Figure 1.46 (e) and (f)) using a purpose-designed hinge made from tubing.

The rear wheel continued to be folded underneath the frame, as in the production model in Figure 1.46. The P2 saw the introduction of castors on the rear luggage rack, on which the bicycle sat when the rear wheel was folded underneath. These too have survived, and can be seen in Figure 1.46.

Unlike the production model, P2's handlebars hinged down, one each side of the package. Also unlike the production model, the seat pillar of P2 consisted of more than one tube which telescoped during folding.

Two more prototypes were built using sliding tubes to produce hinges, this time with 16-inch wheels. Wheel size is a key issue for the designer of a folding bike. Smaller wheels are easier to pack small, but the smaller the wheel the bigger the pothole feels! There is also the 'make or buy' decision to

▼Stiffness and flexing▲

In Block 1 you saw an example of a badly sagging bookshelf. The problem with the shelf was that it was not *stiff* enough: it was deflecting too much under the load applied to it. The stiffer an object is, the less deflection there is when a force is applied to it.

To some degree, the amount of deflection depends on the shape and size of the objects carrying the load. Thus one solution to the problem of the sagging bookshelf is to use a thicker shelf. Greater thickness gives more stiffness.

An alternative is to change to a material which is inherently stiffer. The material's property related to stiffness is called the *Young's modulus*. Two components with identical dimensions will show different stiffnesses if they are made from steel and aluminium, say. The Young's modulus of steel is about three times that of aluminium, so it will make a stiffer component. (A formal definition of Young's modulus will be given shortly.)

SAQ 1.14 (Learning outcome 1.10)

To make the frame of a bicycle light, hollow tubes are used. These are not as stiff as solid tubes. Suggest two ways in which hollow tubes might be made stiffer.

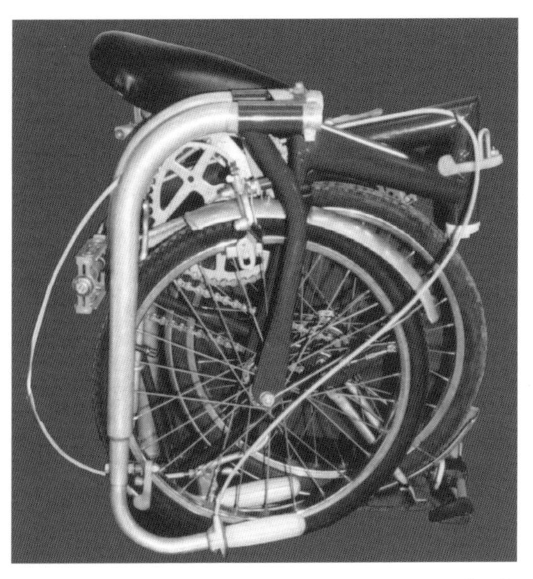

Figure 1. 49 A Brompton prototype. Note the frame hinge on the crossbar

consider. Mass production of bicycle wheels is a big issue; it is much easier for a manufacturer to buy-in wheels produced by a large manufacturer than to dedicate machinery and labour to the production of wheels just for their own product.

Andrew Ritchie's intention was to sell the *design*. To further this ambition, he applied for and obtained a patent in 1981. You will read more about patents in the next block, but for now it is worth noting that his patent may have been difficult to defend, owing to the number of previous designs of folding bicycle that were available. He certainly could not have afforded to defend it if his design had been copied by a large manufacturer, but nonetheless it is a formal statement of the design, a design representation, and a claim to intellectual property.

In total Ritchie built four prototype machines, with a low main tube, between 1975 and 1979, to prove and develop his ideas. His next problem was to turn the design into a product that you or I could buy.

Before pursuing the story I shall look in some detail at the structural design of the bicycle.

6.5 The structural heart of the machine

A bicycle consists essentially of a horizontal beam, to which is attached the wheels and a seat post. It is this beam which, structurally, is the most important part of the bicycle. There are forces acting on this beam when a cyclist simply sits on the machine, and they can be particularly large when the cyclist stands on the pedals going uphill, for example. This beam must provide stiffness for the bicycle: a wobbly bicycle isn't much use because the rider wants the downward force on a pedal to result in work that propels the bicycle forward, not into twisting and bending of the structure. A wobbly frame would also feel unstable to the rider. As has already been noted, the Brompton uses a low horizontal beam. Many bicycles use a high beam, with diagonal posts to join this beam to the pedals etc., in an 'A'-shape (see Figure 1.50). Ladies' bicycles use a double diagonal frame, to reduce the difficulty in mounting the bicycle whilst wearing a skirt (Figure 1.51).

So what is required is a stiff structure that is as light as possible. These two requirements conflict, as reducing weight means less material, which in turn will reduce stiffness. However, it is possible to use the mass available

Figure 1.50 Bicycle with 'A' frame

Figure 1.51 A ladies' bicycle

efficiently or inefficiently. Also, we have this business of choice of materials: aluminium, steel, or something more exotic. Figure 1.52 shows a carbon-fibre composite bicycle. This material has a good stiffness with a low density (so low weight), and in addition the frame is designed to be particularly aerodynamically efficient. A bicycle similar to this, the Lotus Sport bike, was ridden by Chris Boardman when he shaved six seconds off the 4000 metres Individual Pursuit world record at the Olympic Games in Barcelona in 1992.

A racing bicycle like that in Figure 1.52 is built regardless of cost and the suitability of the design for mass production. In our earlier terminology, it occupies a different design space from the folding bicycle, primarily because of the difference in function: to win races, rather than be portable and affordable. Certainly such a bicycle has no requirement to be foldable. The frame of the Lotus Sport pursuit bicycle used by Chris Boardman was moulded from woven sheets of aligned carbon fibre, layered in a mould with epoxy resin, which was then cured. It weighed 8.5 kg. Although the resulting composite has an excellent stiffness-to-weight ratio, weight is relatively unimportant in a pursuit race because only the first 125 m involve acceleration. The rest of the race takes place at a more or less constant speed – as fast as possible. So, it is aerodynamic drag, which accounts for 96% of the

Figure 1.52 Windcheetah monocoque racing bicycle with carbon-fibre frame. This bicycle was designed by Mike Burrows

total resistance to motion, that is the predominant design parameter for a pursuit bicycle. About one-third of the total drag is due to the bicycle.

However, the main criterion for the Brompton is foldability, with weight coming an important second; and aerodynamics are not important at all. A decent ordinary bicycle weighs about 12 kg and is relatively easy to lift and lug about over short distances. Wheels, gears and handlebars need to be mounted on the bicycle; they are available in aluminium and are extraordinarily light because the designs are mature and optimized.

To make a judgement about materials and their use independently of the complexity of a folding bicycle we need to investigate some basic concepts relating to stiffness. As noted earlier, the main structural member of the bicycle is a deep beam onto which the fork, handlebars, seat pillar and chainwheel are attached.

Activity 1.1 Investigating stiffness

Try to find two rulers of similar sizes and thicknesses made from different materials. Wood and plastic will do fine; failing that, a plastic and a steel knife or fork, or something similar. I shall assume you have a wooden and a plastic ruler to hand, that they are the same length and have about the same cross-sectional dimensions. The same dimensions are required because you are going to look at differences between *materials*; if the dimensions change as well then it becomes more difficult to see why any change is occurring.

Bend one of the rulers about both cross-sectional axes. You will find it very easy to bend one way, but not the other. It is hardest when there is more material in the direction along which you are applying the bending force (Figure 1.53). From this you can observe that stiffness depends on geometry.

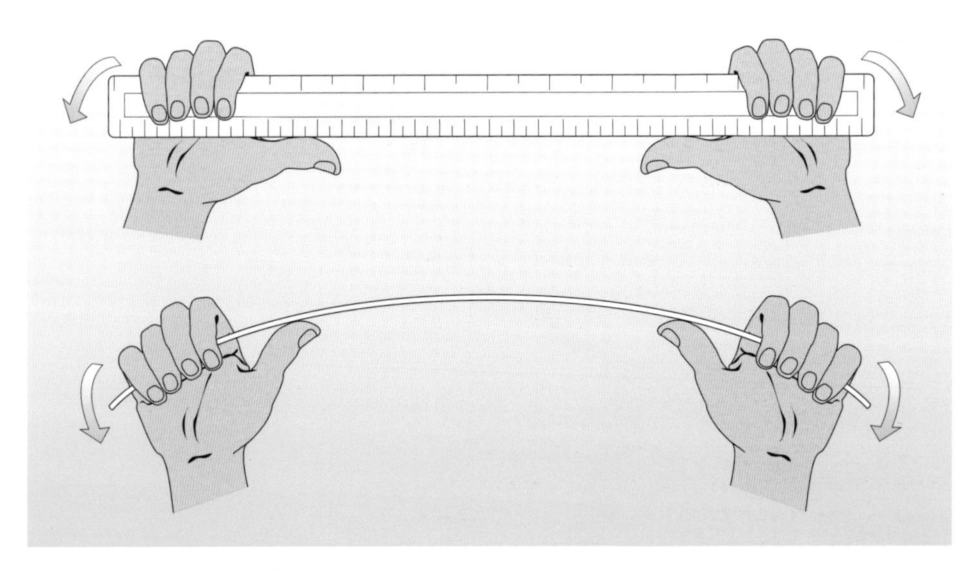

Figure 1.53 Bending rulers

Exercise 1.8

How would the ruler behave if it had a square cross-section?

SAQ 1.15 (Learning outcome 1.11)

Now bend both rulers about the easy way. Which material is the stiffer?

I can now show you these ideas formalized. Figure 1.54 shows the result of pulling on bars made from different materials, each with the same cross-section. Testing of materials is often performed by using tension forces rather than bending, as it makes for a simpler experiment. Bending induces both tension and compression, which can be complex, particularly as some materials show different behaviour under compression than tension.

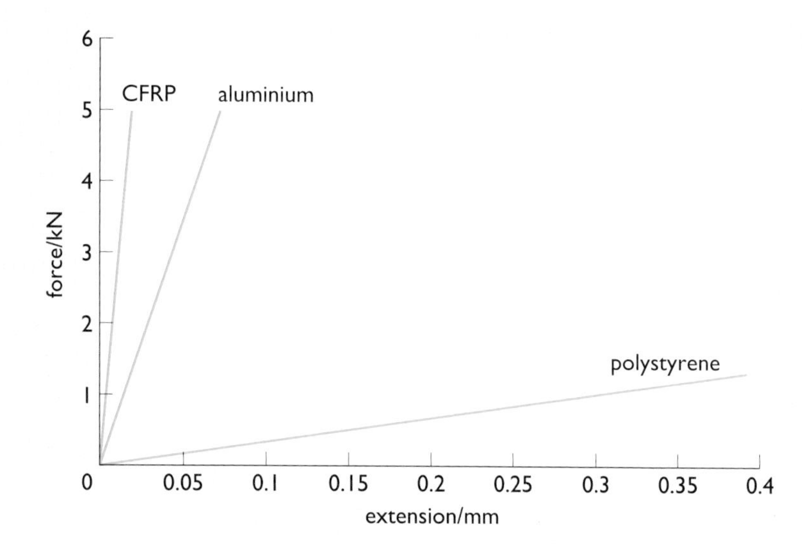

Figure 1.54 The stiffness of bars: polystyrene, aluminium and carbon fibre-reinforced polymer (CFRP)

On the graph in Figure 1.54, the vertical axis is load and the horizontal axis is extension, so the steeper the slope the stiffer the material: there is less deflection for a given load if the material is stiffer. I have been exact in specifying the materials here because different plastics have different stiffnesses, although all are low compared to metals and ceramics.

This is not really a very good experiment. Because the cross-sections are all the same in this case, it is a valid way to compare materials. But the slope of a graph is not a measure of the stiffness of the *materials*, rather it is a measure of the stiffness of a bar of that size made from that particular material. A bar of twice the cross-section would be twice as stiff.

A designer can alter size as well as material to produce a required stiffness, but to make a general comparison of materials I have to show you how to remove the effect of the cross-section, and indeed the length of a particular bar.

First I shall deal with the cross-section. See ▼Materials and stress▲.

My goal is to compare materials independently of their size or shape. Although a thicker rope will bear a higher load than a thinner one of the same material, it will not bear a higher stress before breaking. In fact thick and thin ropes, for example, if they are made of the same material and in the same way, will break at the same stress. So here is one geometry-free measure of a material. The maximum stress a material can withstand is the same irrespective of the shape or size of the sample into which the material is shaped.

A bar stretches under load, and you have seen that differently sized bars made of the same material will stretch by different amounts when subjected to the same load. Again, we want a way of comparing stretches that does not depend on the shapes of the samples we are using. ▼Strain▲ explains how this is done.

▼Materials and stress▲

Strength is an important mechanical property of any material. It is related to how much force can be applied to the material before it fails. Failure generally means fracture: the material breaks into two or more pieces. There are other types of failure, though, such as when a sample is seriously deformed even though it has not actually broken into pieces. Another type of failure is when a sample is degraded so that it can no longer do its job.

Strength is a *materials property*, like its Young's modulus. You cannot change the strength of a material by cutting it into a different shape, for example. You might make it easier to break, but this is not because of a change in the material.

You have already come across force. Whenever you stand on floorboards you are applying a force to them. The loading in that case is relatively complex because of the bending, so let's take a simpler example: a rope used to haul a load.

Given a choice between a thick grade of rope and a thinner grade made of the same material and in the same way, which would you go for if you have to haul some heavy loads?

I expect you would choose the thicker rope. The reason you might give would be that the thicker rope 'looks stronger'. The actual measurable difference is that it can carry a greater force before breaking. Why is this?

Clearly the size of the rope does have a critical bearing on whether or not it will break when it is loaded. A smaller rope can't carry as much force. What we find is that the *area* of rope which is carrying the force is what is important. Twice as much area of material can carry twice as much force.

The force and the area together are used to define what is called the *stress* in the material. The stress is found by dividing the force by the area. It is the stress which controls whether a material will fail. The thick rope will fail at the same stress as the thin rope, even though the force required to attain that stress is much higher in the first case.

We use the Greek letter σ (pronounced 'sigma') to represent stress. F is the force and A is the area over which the force is acting.

Mathematically, we write the definition of stress as:

$$\sigma = \frac{F}{A}$$

The unit of stress will be the units of force divided by the units of area, that is newtons (N) divided by metres squared (m^2). The unit is therefore newtons per metre squared, N/m^2 or $N\ m^{-2}$. (The unit of $1\ N\ m^{-2}$ is sometimes called the pascal, but we will not use this terminology in this course.)

SAQ 1.16 (Learning outcome 1.12)

Two ropes have diameters of 5 mm and 25 mm. What is the stress in each rope if the force applied is 500 N?

Remember: the area which is important is that which the force is transmitted through, which is the cross-sectional area of the rope. This is the circular area you see if you cut a section across (at a right-angle to) the length of the rope and look end-on at the revealed surface.

For revision of areas of circles, review pp. 388–92 in the *Sciences Good Study Guide*.

This brings us to a definition for strength. The strength of a material is the maximum stress that the material can withstand before it fails. As stress varies depending on the area, a smaller piece of material will fail under a smaller force. However, the stress to cause failure should always be the same for a given material.

▼Strain▲

If we apply a tensile force to a material, it will extend in response. This extension is called strain, and is usually barely perceptible, unless you are pulling something like a rubber band.

There is an analogy here with stress. Stress allows us to separate the materials property – its intrinsic strength – from the effect of the size of the sample, which also has a bearing on how big a load the sample can carry. Differently sized samples of the same material will fail at different forces, even though the material has an intrinsic strength that is common to all the components. Similarly, strain allows us to quantify the material's response to loading, independently of the size of the sample used.

Strain is defined as the extension of the sample divided by its original length.

$$\text{strain} = \frac{\text{extension}}{\text{original length}}$$

or

$$\varepsilon = \frac{\Delta l}{l}$$

Strain is represented by the Greek letter ε, called 'epsilon', and the length of the sample by the letter l. The Δ symbol (more Greek: this is the capital letter 'delta') is a shorthand way of saying 'the change in'. So Δl means 'the change in l'. (Δl is said by running the names of the letters together: 'delta el'.) Strain is a measure of the elongation of a material, with the change expressed relative to the original size of the sample rather than in units of length.

In practice, values of strain are usually quite small, and for this reason they are often expressed as percentages. It's easier to say that 'the strain is 0.1 per cent' than 'the strain is 0.001'. To calculate strain directly in percentage,

$$\varepsilon \text{ (in \%)} = \frac{\Delta l}{l} \times 100\%$$

SAQ 1.17 (Learning outcome 1.13)

Calculate the strains, both in absolute terms and as a percentage for the following two examples:

(a) A 10 centimetre bar which is extended by 1 centimetre.

(b) A 100 centimetre bar which is extended by 1 centimetre.

There is a Maths Help section on percentages, on p.339 of the *Sciences Good Study Guide*.

Exercise 1.9

What are the units of strain?

The greater the initial length of a sample, the more it will extend when subjected to a stress. However, whatever the sample's initial length, for a given stress, the strain will be the same. That is, a particular value of stress always results in the same strain, for a given material. This is what makes stress a useful concept.

Exercise 1.10

Figure 1.55 shows three samples of different lengths, all of which are made from the same material. All the samples have the same cross-sectional area (that is, a given applied force will generate the same stress in each sample).

(a) Which strip will extend most when the same force is applied to each sample?

(b) Does the strain vary between the different samples?

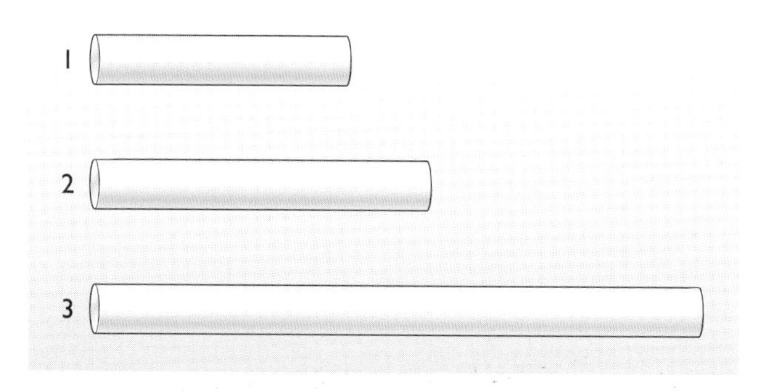

Figure 1.55 Three samples used to measure strain

We now have two measures, stress and strain, which allow us to compare materials independently of the size and shape of any sample we use for testing.

Plotting a graph with stress on the vertical axis and strain on the horizontal axis (Figure 1.56) produces a graph that is similar to Figure 1.54, but generic. That is, the information in the graph is entirely characteristic of the materials; it is independent of the size and shape of the sample used.

The slope of a material's stress–strain graph gives the stiffness of a *material*. This is called its Young's modulus, and is represented by the letter E.

$$E = \frac{\sigma}{\varepsilon} = \frac{Fl}{A\Delta l}$$

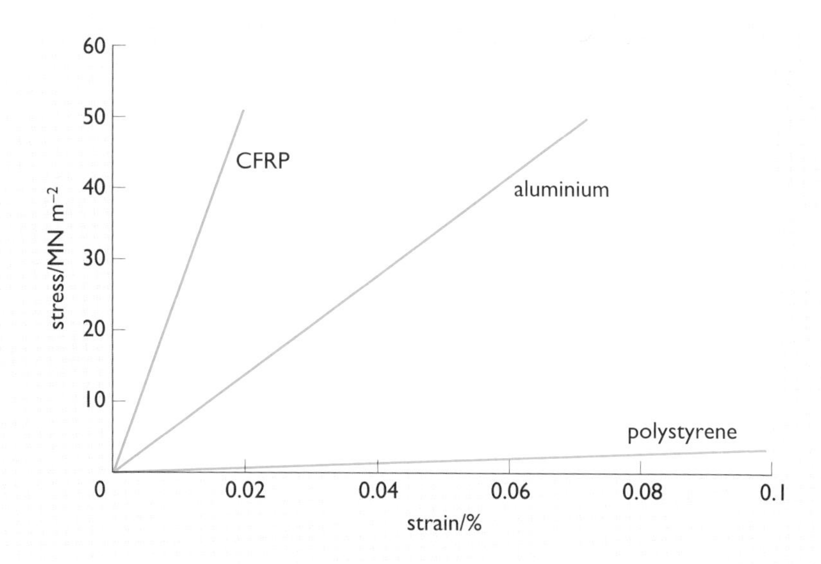

Figure 1.56 The stiffnesses of three materials

So, for example, if we look at the line for CFRP, we can calculate its stiffness by taking the slope of the graph.

The graph is a straight line, and starts at zero. At a stress of 40 MN m^{-2} the strain is 0.015% for the CFRP.

Revision on measuring graph slopes is given in the *Sciences Good Study Guide*, pp. 384–5.

The slope E is therefore given by:

$$E = \frac{40 \times 10^6}{0.015 \times 0.01} \text{ N m}^{-2}$$

$$\approx 270 \text{ GN m}^{-2}$$

(Note the need to convert the strain from per cent to its absolute value.)

So the Young's modulus E of carbon fibre-reinforced polymer is around 270 GN m^{-2}.

SAQ 1.18 (Learning outcome 1.14)

Using the method just presented, calculate the Young's modulus of aluminium from Figure 1.56.

These linear graphs show the behaviour of the materials at quite low stresses only. When the stresses are high enough, particularly for metals, the lines shown in Figure 1.56 can begin to curve (Figure 1.57). If this happens, then the material may be permanently deformed. That is, when the load is removed, the material does not return to its original shape or size. The material has not failed, in the sense that it has not fractured, but the stress has changed its shape. This is what happens when you bend a paperclip.

The capacity of materials to deform permanently is what enables manufacturers to press sheet metal to shape to create, for example, car body panels; but it can be a problem if a structural member may be subject to high stresses in use.

Some types of material do not show the kind of behaviour shown in Figure 1.57. Ceramics, for instance, break when the stress is high enough, without first becoming susceptible to permanent deformation. Think about bending a piece of chalk.

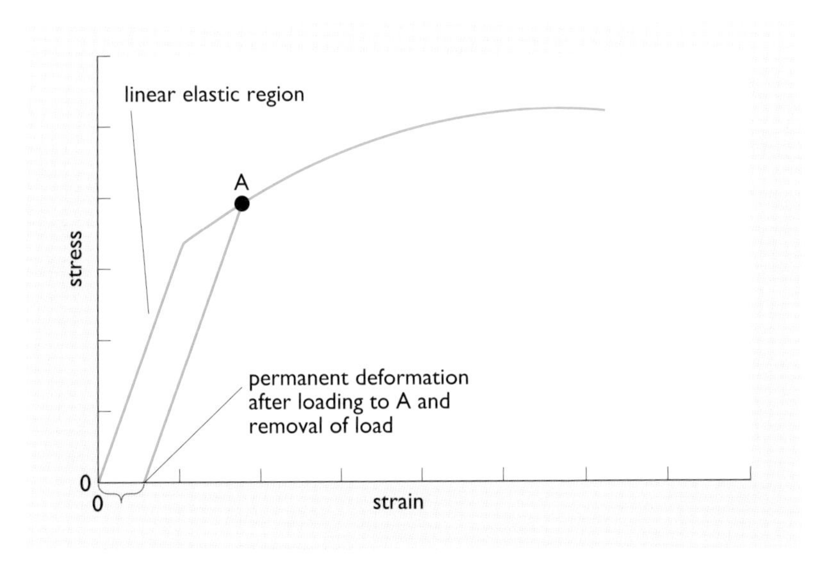

Figure 1.57 Stress–strain curve. If the material is stressed to point A, beyond the linear elastic region, removing the stress takes it down the line parallel to the original linear elastic region. Now when the material is unstressed it has a permanent deformation, or strain, as shown by the brace on the strain axis.

In the portion of the stress–strain graph where the curve is linear, the material is said to be operating in its linear region. When there are no permanent deformations the material is said to be in its elastic region. If both apply, the material is said to be operating in its linear–elastic region. This region is also sometimes referred to as Hookean, a term derived from the name of Robert Hooke, the physicist who first noted this linearity in elastic behaviour,

Not all materials have this useful, linear portion of the stress–strain curve. Rubber for example is non-linear over the whole of its stress–strain curve. Fortunately a surprising number of useful materials show an extensive linear region, which makes their behaviour relatively easy to understand and model mathematically.

Structural materials are used at sufficiently low stresses to ensure that they suffer no permanent deformations during use. It would be most disturbing if one's bicycle wasn't the same shape after the ride to work. In general a material designed for stiffness will be sufficiently strong not to fracture under normal use, although common experience leads us to observe that metals can fracture after repeated use (see ▼Metal fatigue▲).

A steel object loaded to two thirds of the stress which would cause it to deform permanently during normal operation would be considered to be highly loaded. Such a figure provides 'headroom' of one third before something undesirable happens to the material, a safety factor of about 1.5. We are now encouraged to describe a safety factor as a 'reserve factor on load', though you might still prefer to call it a safety factor.

A bicycle designer wants the frame to be stiff, so needs materials with a high Young's modulus. The frame should also be light. You may recall from Block 1 that we use density, ρ, a material property, to give us a way of comparing different materials, in terms of their mass for a fixed volume. Simply comparing the Young's moduli and densities of different metals might not be very helpful because it would be a comparison of properties in isolation. What we need is a method of comparison that takes into account both density and Young's modulus. We need a ▼Merit index▲.

▼Metal fatigue▲

Metal fatigue is an extremely common cause of material failure. Fatigue is a very subtle process, the onset of which can go unnoticed until the fatigued component fails. Fatigue can occur at stresses much below the strength of the material, so may cause failure in a condition that was considered safe by a designer.

Fatigue occurs when the stress in a component oscillates with time. If the oscillations are sufficiently great, they can lead to the initiation and growth of cracks within the material. These cracks can grow until the component fails, often quite catastrophically.

The existence of fatigue has been recognized for over 100 years, but it is only in recent decades that the process has been understood thoroughly to the point where it can be designed against successfully. Failures such as the Markham colliery disaster in 1973, the Hatfield rail crash in 2000, and the loss of three Comet airliners in the 1950s were caused by fatigue. The lessons learnt in each case mean that designers are progressively better equipped to prevent it occurring in future.

▼Merit index▲

A merit index is a combination of certain properties of a material that can be used to inform a process of materials selection for a particular set of criteria. The properties involved will depend on the application. For the bicycle frame discussed here, we are looking for a merit index to find a frame with the best stiffness at lowest weight.

We might also be interested, for the frame, in cost and corrosion resistance, but a merit index will only address one attribute at a time.

In the case of the bicycle, we want to have a high value of E, the Young's modulus, and a low value of ρ, the density. A simple way of producing a merit index is to divide E by ρ: this will give a number which becomes larger as E increases and also larger as ρ decreases.

You will not have to derive merit indices yourself in this course; you will be taught examples of how to use them, though.

A ratio of E/ρ produces a merit index for bars under tension. A high value of Young's modulus and a low density gives a very high merit index (and, of course, the higher the merit index the better). Doubling the Young's modulus whilst also doubling the density would leave the merit index unaltered.

SAQ 1.19 (Learning outcome 1.15)

Table 1.1 gives data for three metals: aluminium, steel and titanium. For each of these, calculate the merit index of E/ρ.

Table 1.1

	Aluminium	*Steel*	*Titanium*
Density ρ/kg m^{-3}	2800	7900	4500
Young's modulus E/GN m^{-2}	70	210	110

The answer to SAQ 1.19 shows that there isn't much to choose between them. Steel's merit index is slightly better than aluminium's and the titanium's is slightly worse. We have to be careful not to extend this analysis beyond its relevance. Remember that this is an analysis of a bar in tension for stiffness. It says nothing about strength.

SAQ 1.20 (Learning outcome 1.15)

We saw in Section 6.6 that the strength of a material is the maximum stress that the material can withstand before it fails. Use the data in Table 1.2 to calculate a merit index for the strength-to-density ratio, σ_f/ρ, for the three metals. I am using the symbol σ_f to represent the stress which causes the material to fail, that is, the material's strength.

Table 1.2

	Aluminium	*Steel*	*Titanium*
Density ρ/kg m^{-3}	2800	7900	4500
Strength σ_f/MN m^{-2}	350	700	300

The titanium used for Table 1.2 is the simplest form of alloy. Titanium alloys can achieve much higher values of strength but are difficult to manufacture. The steel used is one that can easily be drawn into tubes, and the aluminium is a common alloy that is relatively easy to weld.

The answer to SAQ 1.20 indicates that for strength-to-density aluminium shows some advantage, and indeed both aluminium and steel bicycles are available in the shops. One might, however, expect more aluminium bicycles to be available. The reasons that steel is more common are complex, involving cost and ease of manufacture.

Returning to the Brompton, in this case the main structural member of the design is a thin-walled tube acting as a beam – not simply being pulled in tension.

The derivation of merit indices for a beam is quite subtle. It turns out that the merit indices used for a bar (as given above) are appropriate for a beam made from a thin-walled tube only if the radius and weight of the tube are fixed. This clearly is not helpful; we want to have scope for changing the tube radius if it will help. If the radius is allowed to vary, an appropriate index for thin-walled tube of a given weight of material is E/ρ^3.

Exercise 1.11

Calculate the merit indices for stiffness of steel, aluminium and titanium based on this new merit index, E/ρ^3.

If the designer is not constrained by existing dimensions, aluminium starts to show a clear advantage. Early designers with aluminium used the same diameter tube as was used for steel bicycles for their aluminium bicycles so that clips and fittings from the existing market could be used. They were not using the material to its best advantage; more modern designs can be seen using very wide tube.

Merit indices, such as used here, are only part of the story. There are many factors to take into account. Clearly for a designer such as Andrew Ritchie general considerations were of little use. For example, he was designing a bicycle and didn't own a factory. His earliest influence was the Bickerton, which he criticized for being easy to knock about. It is true that because aluminium has a lower strength than steel, it dents more easily. Furthermore, we might expect slenderness to be an advantage for a folding bicycle, and steel is optimized for lower tube thicknesses than aluminium. Steel is ubiquitous for very good reasons.

Finally, before we return to the Brompton story, it is worth thinking a little more closely about the use of symbols, such as F, σ, E and so on, which have featured frequently in the foregoing material. See ▼What's in a symbol?▲.

▼What's in a symbol?▲

When we use a symbol for a physical quantity, such as E for Young's modulus, what does the symbol actually represent?

A symbol such as E represents something more than just a number. Because we are using it as a shorthand for a property (Young's modulus), it must have a unit also. In the case of Young's modulus, the unit is N m^{-2}; so E must represent this unit also. This means that when we assign a value to E, we do so by giving the symbol E both a number and a unit. You can think of this value as being a number multiplied by the unit. For example, the equation

$$E = 9248 \text{ N m}^{-2}$$

is understood as meaning that E has a value of $9248 \times (1 \text{ N m}^{-2})$. Thus it would not be correct to say that Young's modulus was represented by E N m^{-2}, because that would only make sense if E stood for a numerical term alone, such as 9248.

One consequence of this approach relates to the way calculations involving physical quantities are laid out in formal working. Take a calculation involving stress, for instance, which is defined as force divided by cross-sectional area over which the force acts (i.e. $\sigma = F/A$). Suppose we have a force F of 5000 N acting over an area A of 0.005 m^2. We can calculate the stress as follows.

$$\sigma = \frac{F}{A} \qquad (1)$$

$$= \frac{5000 \text{ N}}{0.005 \text{ m}^2} \text{ or } \frac{5000}{0.005} \text{ N m}^{-2} \qquad (2)$$

$$= 10^6 \text{ N m}^{-2} \qquad (3)$$

There are two points to notice here. First, when numerical values are substituted for F and A in line 2, the unit is included, because each of F and A represents the product of a number and a unit. Second, the symbol σ is always equated either to other symbols, as in line 1, or to a product of a number and a unit, as in lines 2 and 3. (Line 2 looks more complicated, but boils down to the product of a number and a unit as the line 3 shows.) Note that in the above example, the correct unit for the answer emerges naturally from the calculation and from the initial data.

The following is not acceptable in formal working.

$$\sigma = \frac{F}{A} \qquad (1a)$$

$$= \frac{5000}{0.005} \qquad (2a)$$

$$= 10^6 \text{ N m}^{-2} \qquad (3a)$$

Here the term in line 2a is just a numerical term with no unit, so it cannot be equated to σ; and in line 3a the unit appears from nowhere. This is rather sloppy, and difficult to follow. It could also to lead to errors. For instance, if the force had originally been stated as 5 kN, it would have been essential to remember to convert it to 5000 N for line 2a. But if the unit is included, as earlier in line 2, it does not matter whether 5000 N or 5 kN is used in the calculation. Either will give a correct answer. That is, line 2 could have been written as:

$$\sigma = \frac{5 \text{ kN}}{0.005 \text{ m}^2} \qquad (2)$$

Now the answer comes out as 10^3 kN m^{-2}, which is still correct.

Consider again the equation

$$E = 9248 \text{ N m}^{-2}$$

If we divide both sides of this equation by 'N m^{-2}', we get:

$$\frac{E}{\text{N m}^{-2}} = 9248$$

The right-hand side of this equation is now legitimately just a number, with no unit. This procedure of 'dividing by the unit' accounts for the style of labelling used for graph axes and tables of data. See for example Figure 1.54 or Table 1.2. The divisions on a graph axis, or the entries in a table, are regarded as pure numbers, so the labelling on the graph shows a physical term divided by the appropriate unit.

Fortunately you do not need to memorize or understand the details of the theory behind these conventions of notation. You can just think of the oblique slash as introducing the unit.

6.6 The first production run

6.6.1 The factory opens

Let's return to the story of the development of the Brompton. Because Andrew Ritchie could not sell his idea, he decided to set up his own factory to manufacture the bicycles. He borrowed money from friends (the interest on the loans was a bicycle) to build 30 bicycles and 20 for sale.

After the increasing complications of the prototyping stage, manufacturing constraints become a powerful influence on the designer:

> Bending one top tube is difficult enough, bending fifty is really tiring.
>
> Ritchie (1999)

The main tube was positioned higher and a simpler bought-in offset hinge replaced the purpose-built tube hinges. This forged hinge, which was critical to the Brompton's development, was from France. So, detailed design changes were made to the hingeing system with some benefits to compactness resulting from positioning the hinge higher in the frame.

The telescoping seat pillar was dropped, becoming a single tube, and the frame was braced with a small diagonal cross-beam to give it extra stiffness.

6.6.2 Batch production

> After the first 50 I got a small-firm government loan to produce batches of 50 bicycles; 400 in a year and a half.
>
> Ritchie (1999)

Designing and manufacturing low-cost tooling was a harder job for Ritchie than designing the bicycle. There are many design routes to producing a lighter frame or a more easily manufactured frame that are not available to a small batch manufacturer. He had to use soft, mild steel for the main tube as he could not bend stronger alloy steel, and the main tube had an aesthetically ugly kink from the bending process. Plastic parts were machined from the solid and metal plate was drilled, cut and bent. The process is essentially scaled-up craft production.

During this period Ritchie learned from customers, and from riding his own bicycle. Indeed, sometimes detailed redesign was necessary in the light of problems and failures brought to his attention by users. The business remained vulnerable, although there was helpful press exposure (see ▼Kew for a ride▲).

6.6.3 Mass production

In 1981 the French manufacturer of forged hinges discontinued production so Ritchie stopped batch production and wrote a Business Plan. After a hiatus of five years, in 1986, Ritchie eventually raised £90 000, half of the money he needed to go into mass production, from a customer, friends and family, and went ahead anyway.

The drive to design for manufacturability continues apace. A tool was designed to curve the main tube, so removing that kink. At the time of writing, 2000, a power press allows the use of a higher specification of steel for the main frame member. The hinges were machined from the solid until expensive forging tools could be bought in 1987. Ritchie is working on removing the skill from the manufacture of hinges.

▼Kew for a ride▲

From *The Standard*, Wednesday 3 February, 1982, p. 19.

At 8.00 a.m. today, as always, Andrew Ritchie arrived at work on his bike. Mr Ritchie works at Kew. He has a workshop there and he built the bike he arrived on in the workshop.

A most remarkable bike it is too. It takes a few seconds to fold it up into a neat package less than 2 ft square which you can pick up and carry anywhere.

No other collapsible bike in the world, says Ritchie, collapses so totally and so easily. And it is just as simple to un-collapse it into a bike again.

Ritchie, an old Harrovian who read engineering at Cambridge, is 35 and says he is appalled by the amount of his life he has already given to this bike.

He had the idea at the beginning of 1976, but it wasn't until early last year that he was able to move into the workshop at Kew and put the bike into production.

He had orders for 30 bikes, mostly from friends and friends of friends. These were made and delivered by last March and, to his great relief, they brought in orders for 20 more.

By the time these were made another 30 orders had come in and there was some welcome help from HMG in the shape of a Small Firms Loan Guarantee.

So this particular small firm stays bravely afloat in these choppy seas, an example to us all.

It currently has a workforce of two – Patrick Mulligan, brazier and Andrew Ritchie, managing director and assembler – and this will increase as orders come in.

Meanwhile there are 56 Bromptons – that is what the bike is called – on the road now and 24 more ready for delivery and I can report that Judge Abdela has been seen arriving at the Old Bailey on one, that Lord Fraser of Tulley-Belton, the Scottish Law Lord, rides one, and that Ritchie's bike, No 7 from the production line, got him from South Kensington to Kew and back all through the blizzards.

It is, of course, an expensive way of making a bike, this, and each one costs £195 by the time you have added VAT. But they are extremely slick little bikes, and with only 80 made so far, think of the rarity value.

A special pedal is used on one side of the bicycle. The pedal folds away, so that it does not project from the folded bicycle. This adds £33 to the current (2001) price list (Figure 1.58). The pedal on the other side does not fold, and nestles in a tangle of spokes and tubes when the bicycle is folded.

The craft route to producing the pedal involves 64 operations. The company bought a piercing and blanking tool for £3500 in 1991 to reduce these operations. All the earlier prototypes and the first batch had a folding crank.

> It was a clever eccentric mechanism that wore easily.
>
> Ritchie (1999)

Figure 1.58 The folding pedal

Figure 1.59 Assembly and testing of the Brompton

It takes 21 minutes to braze the rear frame. The company is investing in an automatic brazing system for the main frame.

> We spend 25 minutes inspecting the bicycle after production and listen to our customer's problems carefully. We have a fatigue rig and keep a constant watch on the details of supplied items. In the early days I sent a £3000 order back to a supplier who did not use a radiused milling cutter.
>
> Ritchie (1999)

6.6.4 Testing

Ritchie cycles regularly, so still tests ideas and changes. During summer 2000 Ritchie was assessing low rolling resistance tyres.

> Gearing was always a problem. To most manufacturers the folding bike is a bottom-of-the-range product. We wanted proper gearing using a bigger-than-average chainwheel, so at various times we became involved in making gears, for example a 13 tooth rear gear. And we had problems with broken teeth. Things are better now. We use the 3 and 5 speed Sturmey Archer hub for gears.
>
> We have resisted the complication of Derailleur gears. All that extra complication is against the philosophy of the design. Perhaps we could sell another thousand bicycles. It's what the market wants. It's a luxury, but we do not respond.
>
> Ritchie (1999)

You can buy a Brompton bicycle, named after The Brompton Oratory, for about £500. Andrew Ritchie won a Queen's award for export achievement in 1995. Not at all bad for a company that, literally, started under the railway arches.

7 Conclusions

7.1 The context of design and innovation

In this final section of Part 1 I would like to lead you to consider several different ways of looking at design. This is meant to introduce some of the issues and debates which have engaged designers.

We have examined several examples of design. We have seen that design is a complex activity. It has many stages and at each stage must take into account both opportunities and constraints. Designers try to speculate as much as possible within the constraints of time and cost as well as the requirements of customers and clients. The activity or process of design is always treading a fine line between freedom and constraint. Getting the balance right seems to yield useful and satisfactory designs.

There does not seem to be any recipe for achieving this balance. Design is complex with many factors to take into account. There are models of the design process which can act as useful guides to stages and outputs in this process. But they will not tell you how to design a particular thing.

7.2 Innovation

Design is rather like problem solving. We try to define the problem, perhaps in terms of a client's requirements, then search possible solutions for a satisfactory outcome. However, design itself is much more difficult to define. The design problem, although specified by requirements, acquires new constraints as the design proceeds. New possibilities and new needs are continually suggested as the design is developed from concept to detail and onto market. The problem does not remain static.

A major resource for design is technology, and technology may include principles of engineering and applied science, or more tangible products of science such as new materials or electronic devices available for use in new designs. The resources are specific to a particular problem. Any problem has its own context, which might consist of market, customer requirements, or ways of working within a particular industry.

With technology and contexts we can classify several types of design. First, consider a well established context, such as personal transport in some form of automobile to run along roads. Problems of pollution require new means of propulsion. These might be solved using new technology. This may not be completely new technology but rather technology new to the context. So a fuel cell developed for space applications may be applied to cars.

Second, consider a well established context and the development of technology already used in that context. An example of this might be developing more efficient internal combustion engines for cars.

Third, a new context arises from social and cultural trends or scientific discovery. For example, the new context of popular long-distance air travel. This new context has led to the design of new aircraft with some new technologies. However, for the most part it is a question of using well established technologies.

Fourth, a new context might combine with new technology. Examples might be the development of radar or nuclear weapons during the Second World War. More benign examples include the new responses to home energy use following the rising costs of fossil fuels, or the use of autonomous vehicles to explore the sea bed or maintain offshore oil and gas production facilities.

These four classes may be arranged as a table (Table 1.3). Names are given to each of the classes. Innovation can be viewed as including something new in technology or in the context to which it is applied. Inventions are not always closely attached to context. However, these inventions are applied science rather than design if no specific product emerges. When a new technology is matched to a new problem or context a design invention can emerge.

Table 1.3 Technology in context used to distinguish types of design

	Old technology	*New technology*
New context	innovation	invention
Old context	routine design	innovation

We should be rather generous in our interpretation of 'technology'. It will include new ideas or principles on which designs can be based. These new technologies do not always deliver new products easily. Turning out a product that works is usually fraught with difficulties.

The world of large structures provides good examples. Some of the cathedrals we see today are the survivors of designs produced by secret guilds of masons seven hundred years ago. Flying buttresses (Figure 1.60) were a design innovation that looked beautiful and allowed tall walls with large windows to be built.

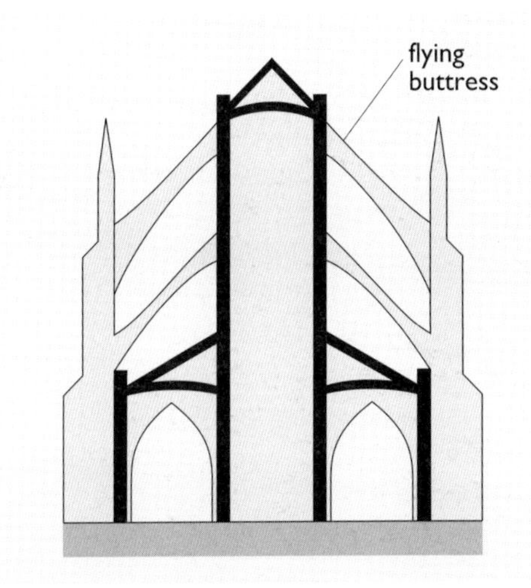

Figure 1.60 Flying buttresses

The function of the flying buttresses, and the decorated finials on the top of the verticals, is to keep the internal forces inside the stone so that all elements are in compression, or pushing against one another. Stone is very good at resisting compression but very poor at resisting tension forces. The simple arch also makes use of this property of stone to span spaces in bridges and vaults. There were other ways of achieving the same artistic effect of tall walls and large windows. Iron reinforcing bars were placed in the walls to resist sideways forces. In some mediaeval cathedrals you can see both methods used, literally side by side. On the outside, public, wall of the church there are buttresses and plenty of ostentation but on the interior, private, side there are fewer, less ostentatious buttresses and more iron reinforcing within the walls. It is informative to note that parts of cathedrals did fall down. This contributed valuable knowledge to subsequent designers. It gives us a valuable message as well. Designers learn a lot from failures.

The first large suspension bridges in China, several hundred years ago, were using the same principles as the bridges built by Telford and Brunel in the 1800s. These later engineers were rapidly pushing to the limits the available

Figure 1.61 Brunel's Clifton suspension bridge, painted by Samuel R. W. S. Jackson (1794–1868)

materials and the scientific understanding of their day. There was a great deal of uncertainty in their designs, although they had the benefit of some scientific analysis and could build simple theoretical models of how structures might behave. Brunel, for example, applied elements of numerical modelling to the business of building bridges.

Section 1.9 defined design as the process of converting generalised ideas into specific plans. So designing is a process, possibly shared processes, by which we change things in the world. This is what Brunel was doing when he created the new form of a chain-link, large span, suspension bridge across the River Avon at Clifton in Bristol (Figure 1.61).

Telford and Brunel were in competition to build the Clifton suspension bridge. Telford's design used two towers built up from the river's bed (Figure 1.62). Brunel commented sarcastically that he had not thought of building towers from the sand of the river bed when there were good rock buttresses to build on.

Figure 1.62 Telford's design for the Clifton bridge

Telford's design was more cautious than Brunel's, so the towers had to be closer together because the span was more limited. The cost of Telford's towers was much greater, but the structural uncertainty in his design was less. Put another way, it had a higher safety factor. Nowadays the analysis of a suspension bridge is routine; the relationships between the cost of towers and the tensile strength (that is, strength under tension or pulling) of the main wire ropes are well understood. The properties of the relevant materials are known in great detail.

However, there is still uncertainty. The Millennium suspension footbridge over the Thames was found to behave in erratic ways when opened to large numbers of pedestrians in June 2000. The swaying bridge was alarming to use and was closed for repairs within days.

Interestingly, Brunel was a financial disaster for people prepared to invest in his designs. The Clifton suspension bridge was not finished in his lifetime because of a shortage of money. Design is not necessarily profitable; it is a risky and uncertain business.

7.3 Uncertainty

We have noticed during the block that designing takes place under various degrees of uncertainty. This is another way to classify design. From the early stages to a concept are full of uncertainty, whereas later stages can be more routine. However, in all design projects it is not known how a design will perform until it is completed, tested and then used. Each design project is a response to a new situation. It would not be design otherwise. So designers face uncertainty in all they do. They try and reduce uncertainty by:

(a) using models to predict how designs will behave;

(b) using experience gained from the performance of previous designs for similar problems.

As experience (of success and failure) increases and predictive models become more accurate, the inherent uncertainty in design decreases. A well-established technology which is matched to a well-established context has little uncertainty. In these circumstances designers have the task of creating variations and modifications on the basic design.

Once a design space is well-understood, the production of variant designs becomes a mature business, where ingenuity goes into making the processes as efficient as possible. *Managing* design becomes more important than the fundamental activities of innovation and design.

In the building industry the creation of a new McDonalds restaurant uses the same well-established rules all over the world. Similarly, some types of automotive and electronics factories have become almost standard items. The machines and assembly lines can be established from a greenfield site in two years. These types of design are standard and routine. Design uncertainties are low but other uncertainties of markets and competition remain.

However, large international companies retain a mix of innovative products and variant designs. They spread their risk across many products, recognizing that they have to innovate to survive. Today's innovative products are the basis of tomorrow's variant designs. A little later we will see that televisions are an example of variant design, yet their manufacturers are also creating new products such as camcorders, digital cameras and portable DVD players.

SAQ 1.21 (Learning outcomes 1.4 and 1.8)

Looking back over the examples of design that we have considered in the block, identify sources of uncertainty in design projects.

▼Culture of innovation▲

The following is by Tim Brown (European Director of Ideo, a product development company), and was published in the *Financial Times*, 17 November 1997.

Innovation requires, above all else, a willingness to embrace chaos. It means giving free rein to people who are opinionated, wilful and delight in challenging the rules. It demands a loose management structure that does not isolate people in departments or on the rungs of a ladder. It needs flexible work spaces that encourage a cross-fertilization of ideas. And it requires risk-taking.

Yet if innovation has become an over-used buzzword, it is only because we all recognize innovation as a competitive weapon, a necessary component for future success.

In the world of product development, where clients originally turned to external consultants to provide additional capacity, speed or a particular technical expertise, we now see them looking for guidance on how to innovate. They are looking for a process.

The response has to be that innovation is not something prescriptive. You can't legislate for it. Rather, it is something organic, something that grows and is nurtured, usually from the bottom of the organization up. Experience shows us that innovative cultures usually begin with a tangible project the success of which gives birth to another and another. Such projects or definable goals are also the elements which keep the fun and freedom from degenerating into non-productive anarchy. Moreover, if you want the combined workforce behind you, but take away their desks, their titles and their personal power bases, you must give them something in return. Job satisfaction is a great motivator.

In a culture of innovation, enlightened trial and error beats careful planning, and risk becomes an essential part of the process. Rapid prototyping means that you can evaluate a concept before you have invested too much in it. It also gives the participants the stimulation of seeing their ideas put into practice and sustains their enthusiasm when more concrete rewards are less evident.

Early prototyping and testing also allow you to fail and if you are not failing often, then you are probably not risking enough. So, having said you cannot legislate for innovation, here are some common themes which give some ideas on how to get started:

- Treat life as an experiment – constantly explore new ideas through projects.
- Innovation is a team sport – be smart about creating and sustaining hot groups of energetic, opinionated people.
- Risk a little, gain a lot – fail quickly and often by user testing and prototyping ideas.
- Identify goals – build multi-disciplinary teams then give them a common aim to create products or services grown of collaboration not compromise.
- Observe the consumer, don't ask him what he wants – find ways to get under the skin of the end-user to identify new needs, opportunities and possibilities.
- Allow serendipity to play its part.
- Space is the last frontier – provide environments.

Innovation has been identified as a critical component in business success. However, innovation involves uncertainty and risk. The imperatives for companies to move away from the routine to new contexts and new technologies are now very strong. However, the tendencies of many designers are to reduce uncertainty. They tend to be more like Telford than Brunel in the case of the Clifton suspension bridge. As we have discovered, designers cannot escape uncertainty but that does not stop them trying to minimize it where possible. To maintain a balance between staying within the bounds of known and well understood designs and exploring new possibilities companies try to create a ▼Culture of innovation▲.

In discussing technology, innovation and uncertainty we have concentrated on the functional or engineering performance of designs. However, there are other important features of any design, such as style. This is how a design appears to a customer. Style can be a major factor in the commercial success of the design.

7.4 Style

The term 'industrial design' is often used to denote those design activities mainly concerned with the appearance and aesthetics of a product. In contrast the term 'engineering design' is often used narrowly to indicate those design activities which deliver the physical performance of the product. Thus the *engineering* design of a Sony Walkman is about the mechanisms of tape or CD movement and the electronics to decode and play the music. The *industrial* design is about the appearance of the box, the ease of loading and unloading and the layout of controls. Both are critical to the successful product.

Recalling the plastic kettle, the style of the jug kettle was a major factor in its success. However, the technical problems of plastic materials and manufacture in jug form are also considerable. Their solution was needed for a successful product.

In some cases the technical issues of engineering design can be separated from the style issues of industrial design. This is particularly the case with mature products where functional development is limited by the technology. The functional parts of these products can be clothed in a new aesthetic.

Take for example, a television set based on the standard cathode-ray tube technology. (A cathode-ray tube, or CRT, is the part of a conventional television set which creates the pictures. The television screen forms one end of the cathode-ray tube.) The technology of standard television sets is well-established. Further, the manufacturing processes, such as populating printed circuit boards (that is placing the components on the board and soldering the electrical connections) is remarkably well-established and standard throughout the world and across competitors.

The housing of a televison set is a large, plastic injection moulding that can be changed in shape, subject to considerable limitations imposed by the shape of the tube. Figure 1.63 shows an interesting extreme of a style from the 1970s being used to house a television set made in 2000 and using the technology of 2000.

Figure 1.63 Television style from 1970s sold in 2000

Typically a company such as Sony will cater for changing tastes in style across its markets by engaging in industrial design separately from engineering design, or by seeking expertise on appropriate style through design consultancies in different markets and countries.

The judgements that designers and consumers make about the balance of style and function are subjective. That is, people make their own judgements which can differ widely from individual to individual. For example, a general-purpose bicycle is designed to be quite stylish. The technology is standard, and style is the main differentiating factor between different manufacturers.

In contrast, a mountain bike may be predominantly functional. However, I don't think many people would disagree that a mountain bike is designed to be stylish. Many of the features are not only functional. You could argue that the mountain bike is full of style with a corresponding high cost. A mountain bike might feature chromium plating, disc braking, roller bearings, and advanced suspension. Not all mountain bikes have these features and their functionality is debatable. The bicycle would work perfectly well with lower specification parts, so it could be argued that these features are really part of the style associated with this type of product.

Designed products can be seen and thought about in many ways. They are invested with 'life' and meaning by people who use them. This extra meaning, for example in the fashion status of functional products (such as catering equipment now popular in domestic kitchens), is one of the key elements in a successful product. Designed objects acquire many layers of meaning. The process often starts with designers themselves and is continued by advertising. This interpretation is what people do well. In fact it is one of the important activities in designing where new possibilities for developing partially completed designs emerge. The design evolves and adapts as the designer sees and thinks in different ways.

7.5 Examples of context: televisions, aircraft and soap powder

Designs are not just differentiated by what they do and how they do it, or what they look like, but also by the wider social and economic contexts in which they are created and used. To illustrate this, let us return to the design of television sets.

Traditional television manufacture is a global industry, and a cluster of companies, including Sony and their suppliers, are located in South Wales at the time of writing (2000). This cluster was the result of considerable inward investment several years ago and is now under threat from new flat-screen technology (see ▼Gas plasmas and electron guns▲).

The South Wales companies concentrate on the production of a traditional product and incremental modifications, such as replacing analogue electronics with digital electronics and changing the aspect ratio (width-to-height proportion) of the screens. The manufacturing plant is terrifyingly efficient. Sony maintains its niche by manufacturing and distributing at minimum cost. To do this it needs to control its supply chain and both influence and respond to its customers. The production of the CRT is costly, requiring expensive, dedicated machinery. To maintain a dominant position, Sony outsources the very large plastic injection mouldings for the cases of

▼Gas plasmas and electron guns▲

Flat-screen televisions use a completely different technology from the conventional cathode-ray tube sets which have been around, relatively unchanged excepting the advent of colour, since the 1940s.

In cathode-ray tube televisions, the image is made by firing a beam of electrons (emitted by a *cathode*, hence the name) at the screen, which is coated with a phosphor layer that glows when it is hit by the beam. The beam scans across the screen in lines, with the intensity varying constantly to build up the picture. The beam is scanned 25 times a second, so the image appears constant. The reason that CRT-based televisions are quite deep is that there must be room for the beam to be pulled up and down (by electromagnets) to the top and bottom of the screen after leaving the cathode.

Since the late 1990s flat screens have become available (and I am not talking here about 'flatter tubes', which are still just CRTs). These flat screens are based on tiny glowing plasma cells. The screen is made up of many hundreds of thousands of these cells. The television can be made as flat as the associated electronics will allow, hence televisions with these screens are much thinner than those using CRTs. They are probably suitable for hanging directly onto a wall.

televisions. These injection moulding companies are in the region because of the concentration of inward-investing global companies such as Sony.

Interestingly these moulding companies work differently from traditional injection moulders, called 'trade moulders' in the UK. These trade moulders were established by engineers with toolmaking abilities. They were supported by the plastics expertise of large chemical companies such as ICI. Their particular expertise lies in designing tools that can be filled with molten plastic, efficiently creating mouldings which are dimensionally stable across hundreds of thousands of repetitions. A product designer can go to these companies with a shape and rely on them to produce it to a specification. Sony, however, separate tool design from the moulding process, so maintaining a more detailed control over its suppliers. The moulding companies are all within a day's delivery of the assembly plant. The context of the design of the television case is thus one which allows Sony to maintain full control over its production, including all suppliers. Sony wants to reduce its uncertainties and risks in a market which is highly competitive.

Exercise 1.12

How are designs and designing in Sony responding to competitive pressures?

Sony represents a context for design which is driven by a mature competitive market, and which is configured to let Sony have better control over its production than do the competitors.

Other contexts for design are the ways an industry is structured, its markets or even politics. Let us look at the aircraft industry which includes airframe designers such as Boeing or Airbus and engine designers such as Rolls-Royce or General Electric.

In the aircraft industry the life of a design and its variants is of the order of 40 or 50 years. Rolls-Royce maintains a full range of jet engines, some used in aircraft built in the 1960s. There are a number of key technologies, business relationships and supplier industries that come together to make an aircraft: electrical and hydraulic systems, structures, materials, aerodynamics and engines. Figure 1.64 shows the dominating design requirements at different points in the airframe.

As discussed in Section 6.5, fatigue is a very important concept in designing for the operating life of a product. The extent to which fatigue will occur in a

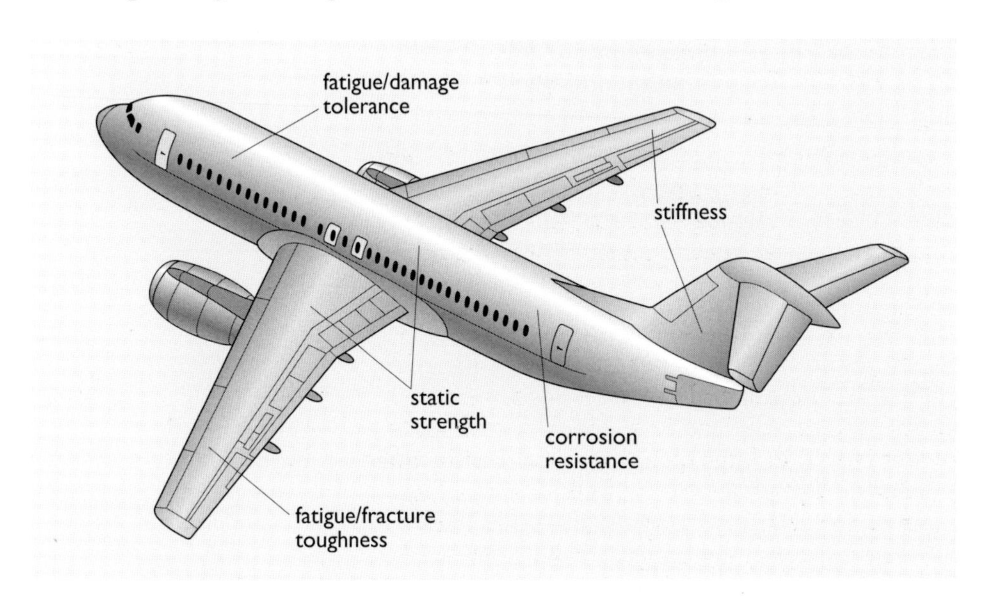

Figure 1.64 Dominating design requirements for an airframe

particular design depends on how the product is used – the operating conditions. Unfortunately, detailed understanding of the long term effects of this fatigue is difficult to predict without extensive testing. A lot of the knowledge gained about fatigue is empirical. It is case-by-case and material-by-material. Patterns do emerge but experience shows that they must be interpreted with caution. For the long-lifetime products such as aircraft the ability to predict the effects of operating conditions and régimes is critical. I emphasize this because a lot of our perceptions of good design hinges on quality and more particularly reliability. Designers need to consider not only the static function of a product but how it is used – the typical operating cycle.

How are the design activities of airframe companies dependent on context? Consider materials development in the context of a company like Airbus. The industry, generally, is a user of advanced aluminium alloys because they are strong for their weight. This is clearly of importance for aircraft. Associated with each alloy is a large body of knowledge on performance from tests and in-service data. To develop and use a new alloy is a significant commitment.

Airbus is interested in new alloys such as aluminium–lithium alloy for use in airframes. Lithium is a light element (it has a low density), so if an aluminium–lithum alloy can be developed with comparable properties to existing alloys, then it will reduce the weight of the airframe. The consequent effect to the airline in reduced operating costs or increasing payload could be a major commercial benefit. However, Airbus does not develop alloys. They are developed and promoted by alloy producers, so the product-design company works closely with the metal-design company, which also works with Airbus's competitors. The context of design becomes more and more intricate.

To add to this complexity, if a competitor such as Boeing is working on such an alloy then Airbus cannot afford not to parallel the work, even if they know that the prospects of success are remote. The penalty of failure is failure of the company. Designers in the different companies watch each other, not in the sense of industrial espionage, but in the open forums where companies that simultaneously compete and collaborate meet. The same principle applies to aeroengine companies. A new alloy that will operate at higher temperatures makes a more efficient engine, so if General Electric is working on a new class of material then Rolls-Royce also has to work on this class.

Unfortunately the ways that these complex sets of interrelationships behave changes from business to business. So what may be true of the aircraft industry cannot be generalized to the computer industry, or to telecommunications or software, for example. Furthermore the complexity of the interrelationships in a particular industry is determined not only by the technical issues, such as those discussed above, but also by broader economic issues.

In order to provide a contrast, compare the aerospace context with the design of packaging for household foodstuffs and cleaning products. Unilever and Procter & Gamble compete fiercely in a global market for soap powder. They try out new packaging in test markets and watch each other's test markets carefully.

In the 1990s plastic flexible bags of soap powder (the usual pack having been a cardboard box that stood upright on the sink) were introduced to one of the test markets. These sold well in the first month and maintained sales for a second month. So both companies became involved in a race to offer flexible packs world wide. As soap powder is sold in large quantities world wide, this required a considerable design effort to equip factories with the new tooling for the new packaging. The advantage to be gained in the market critically depended on the schedule for introduction of new designs of tooling to produce the machines to make the new packages. Although complex packaging requires a longer lead time in the setting up of a production line, the time scale involved is nevertheless a matter of weeks and months, rather than years. Thus 'design' operates quite differently in the world of soap powder from the world of airframes. Designers of soap powder can try out possibilities in markets, but aircraft designers only get one shot.

7.6 End note

To end on a lighter note I would like to touch on a famous design case study from the history of navigation. Clocks in the 1700s only worked accurately if their mechanisms were kept still on the mantelpiece. But clocks were needed to tell the time at sea for the determination of longitude. Valuable prizes were offered and eventually a satisfactory design was created by John Harrison. The story of this design and its social, political and economic contexts is related by Dava Sobel in her book *Longitude* (Sobel, 1995).

The new design certainly met a need. But the design did not behave in new ways – it still counted the seconds mechanically. There were numerous innovations in the internal clock mechanism to achieve the accuracy that was required. The new design behaved just as a clock should behave but in new conditions. The matching of design and operating conditions was the key to successful design. The design of the marine chronometer allowed the mechanisms of the clock to work independently of the disturbances of a voyage at sea. Clever design of the clock mechanism decoupled the internal function from external context.

Design is complex and there are many ways to handle this complexity. Design problems are commonly broken down into stages of increasing detail and definition. There are many activities in the process of design, and these are used in different mixes and with different emphases according to context. The various ways that designers approach complex problems where there are no clear or even rational answers, under conditions of extreme uncertainty, makes design an exciting area of study. Design lies at the heart of engineering practice – it makes things which are useful, beautiful and, perhaps more often than it should, unsatisfactory. But that is what happens when we engage with complexity – surprise and disappointment, success and failure.

SAQ 1.22 (Learning outcome 1.10)

From the examples and case studies described in this block pick out what you think are informative examples of:

(a) innovation

(b) uncertainty

(c) style

(d) context.

Make sure that you have a good reason for each choice!

This concludes our look at the process of design. In the rest of the course we'll look at many of the constraints on engineering within which designers have to work, and you will see illustrations of many different products and applications. The examples will be used to illustrate particular engineering principles, but in all cases try to think of them in terms of the overall design: context, function, style and innovation.

8 Learning outcomes

Having studied the block you should be able to:

1.1 Recognize that functional artefacts have had input from a designer, with greater and lesser degrees of engineering input.

1.2 Identify that engineering designers work within constraints of finance, materials properties, desired functionality, human factors, etc.

1.3 Understand that design exploits models of the product being designed, whether those models are physical mock-ups, computer-based models, or mathematical models which explore an element of the product's performance.

1.4 Understand that there is rarely a unique solution to any design problem. Part of the skill of a designer is in finding a problem–solution pair, and the best compromise.

1.5 Understand how models of the design process are formulated, and how they can be applied to understand the development of a particular product or product family.

1.6 Understand that models of the design process, while useful, cannot guarantee good design or provide a template by which all designs can be judged.

1.7 Understand that early choices about design can have large influences on the available final solutions.

1.8 Appreciate the steps required to move from a conceptual design to a functional product within a process of innovation.

1.9 Critically evaluate the success of a designed product, and suggest concepts for improvement where necessary.

1.10 Understand design-related terminology such as innovation, context, uncertainty and style.

1.11 Understand the concept of stiffness, and that the stiffness of a component can be altered by changing its dimensions or changing the material from which it is made.

1.12 Understand the concept of stress, and how it can be calculated simply from the force and the cross-sectional area.

1.13 Understand the concept of strain, and how to calculate it.

1.14 Understand the concept of Young's modulus and how to calculate it. Be able to distinguish between Young's modulus as a material's property and stiffness as a component property.

1.15 Understand the principle of a merit index for comparing different materials, and be able to perform simple calculations of merit indices.

Answers to exercises

Exercise 1.1

(a) The shape of the Futura was the same as for the conventional kettles; the polymers allowed brighter colours to be used, though.

(b) The polymer kettle was cheaper.

(c) The heating element had a lower power, for safety reasons, so the water took longer to boil.

(d) The polymer used in the Futura gave the water an odd taste.

(e) The polymer had a shorter lifespan, as the colours faded and the kettle began to look shoddy.

Exercise 1.2

You may have offered any of the points shown in Table 1.4. The items in my list are only suggestions: you may not agree with many of them. The important thing we're looking at is your *perception* of the product.

Table 1.4

Good points	*Bad points*
Cheap to purchase	Difficult to clean
Easy to fill	Hot to hold or steam escapes onto hand
Easy to pour	Difficult to grip
Easy to judge the amount of water in the kettle	Heavy
Easy to clean	Difficult to fill
Available in colours which suit my environment (home or work)	Difficult to pour
Can be repaired if necessary	Difficult to judge the amount of water in the kettle
Stable	Expensive
Nice to have on display	Difficult or impossible to repair
Seems safe	Poor image – I don't like it on display
	Unstable
	Seems unsafe

Exercise 1.3

French's model includes loops to earlier parts of the design process, allowing the designer to 'go round the design cycle'.

Exercise 1.4

While there was an established need for *kettles* there was no perceived need for a *plastic* kettle. The original, unsuccessful plastic kettle, was someone's new idea. So, in this respect there was no need or task to be performed.

However, once the company decided they should make a plastic kettle (in this case it was the idea of the Technical Director), the need for a plastic kettle is established. The task for the designer was to design one. So in this respect the models are applicable.

Exercise 1.5

(a) In kilometres per hour, the average speed is:

$$\frac{\text{distance}}{\text{time}} = \frac{38\ \text{km}}{(169 \div 60)\,\text{h}} = 13.5\ \text{km h}^{-1}$$

(b) In metres per second, the average speed is:

$$\frac{\text{distance}}{\text{time}} = \frac{38\,000\text{ m}}{(169 \times 60)\text{s}} = 3.75\text{ m s}^{-1}$$

Exercise 1.6

For Puffin:

$$P_{\text{Puffin}} = C_{\text{Puffin}} \times x$$

First we need to calculate the value of x.

There is help on using formulae on p. 365 of the *Sciences Good Study Guide.*

$$x = \frac{M^{3/2}}{\sqrt{S}}$$

where

M = the total mass of the plane and the pilot in kg;

S = the cross-sectional area of the wing.

For Puffin,

$$M = 63.6\text{ kg (the mass of the plane)} + 61\text{ kg (the assumed mass of the pilot)}$$
$$= 124.6\text{ kg}$$
$$S = 36\text{ m}^2\ \text{(given on p. 40)}$$

Inserting these values into the formula for x gives:

You can find additional help on powers on p. 353 of the *Sciences Good Study Guide.*

$$x = \frac{[124.6\text{ kg}]^{3/2}}{\sqrt{36\text{ m}^2}}$$

which can be rewritten as

$$x = \frac{[124.6]^{3/2}\text{ kg}^{3/2}}{\sqrt{36}\sqrt{\text{m}^2}}$$
$$= \frac{1389\text{ kg}^{3/2}}{6\text{ m}}$$
$$= 231.5\text{ kg}^{3/2}\text{ m}^{-1}$$

Since

$$P_{\text{Puffin}} = C_{\text{Puffin}} \times x$$

$$P_{\text{Puffin}} = C_{\text{Puffin}} \times 231.5\text{ kg}^{3/2}\text{ m}^{-1}$$

For Condor:

$$P_{\text{Condor}} = C_{\text{Condor}} \times x$$

Calculating x

$$M = 34\text{ kg (the mass of the plane)} + 61\text{ kg (the assumed mass of the pilot)}$$
$$= 95\text{ kg}$$
$$S = 75\text{ m}^2\ \text{(given on p. 40)}$$

$$x = \frac{[95\text{ kg}]^{3/2}}{\sqrt{75\text{ m}^2}}$$
$$= \frac{926\text{ kg}^{3/2}}{8.7\text{ m}}$$
$$= 106.5\text{ kg}^{3/2}\text{ m}^{-1}$$

Since

$$P_{\text{Condor}} = C_{\text{Condor}} \times X$$

$$P_{\text{Condor}} = C_{\text{Condor}} \times 106.5 \text{ kg}^{3/2} \text{ m}^{-1}$$

Exercise 1.7

My suggestions are:

1 There are likely to be many choices for the details.

2 Constructing models (probably computer models) of the design and studying the behaviour of the models is likely to be time consuming.

3 Testing and evaluating possible designs to find a best or a satisfactory solution is laborious. Note that finding a best design is very difficult and time consuming in complex problems. The costs in time and money of small improvements in performance can be large when we are close to a good solution.

Exercise 1.8

If the ruler had a square cross-section, it would have the same stiffness regardless of which side was chosen to bend it: the distance would be the same in both cases.

Exercise 1.9

Since strain is extension (a length, measured in metres) divided by the original length (also a length, measured in metres), it has no units. The units cancel when one is divided by the other.

Exercise 1.10

(a) The longest strip will extend the most, because larger extensions are generated in longer samples, for a given applied stress. Thus the shortest sample will extend least.

(b) The strain is the same in each strip, even though the extension is different. The stress is the same in each sample, so this will produce the same strain, as the same material is used for each.

Exercise 1.11

Table 1.5

	Aluminium	*Steel*	*Titanium*
Density ρ /kg m^{-3}	2800	7900	4500
Young's modulus E/GN m^{-2}	70	210	110
Merit index $(E \div \rho^3)$/N m^7 kg^{-3}	3.2	0.42	1.2

Exercise 1.12

There are several points which you could have picked out:

(a) Variant designs are produced in high volumes, so maximizing use of available capacity.

(b) New cases are produced by suppliers in a highly controlled way so that production plans and schedules are not disrupted. This is critical for high-volume production.

(c) There is a mixture of variant and innovative designs.

(d) Existing televisions are regularly re-styled to ensure that they remain popular in the market.

Answers to self-assessment questions

SAQ 1.1

Although I don't know what is in your particular room, I can help you answer the question by illustration.

First, looking at the question overall, any human-made object you choose *must* have been designed. It did not arrive in its form by accident. Furthermore, every part of it was designed, and is there for a reason. I shall now discuss some specific examples. You should have had similar thoughts for whatever examples you chose.

Let us take my coffee mug. The designer had to ensure that it could hold hot liquids without disintegrating, and without it being too hot to pick up. It also had to look attractive, and not be too expensive. There are probably several other things that I did not immediately think of. For example, it must be comfortable to hold even when full of liquid and therefore quite heavy.

Nevertheless, it is a fairly standard ceramic mug (no innovation, although it is decorated in an unusual way). It works very well. The material it is made of is strong enough to contain the weight of liquid. The mug has a handle, which enables me to pick it up without burning my fingers. The ceramic material is not a good conductor of heat, which helps prevent the handle getting too hot, at least during the normal length of time for which the mug contains a hot liquid. A drawback of this material is that it is brittle, and if I drop it, the mug will probably break. The mug has rounded edges and is not dangerous (although it might be if I dropped it).

As another example, consider my chairs. I have two in this room. The designers of the chairs had the problems of giving me something comfortable to sit on, something reasonably portable and mobile, and making it reasonably cheap. Both chairs are quite ordinary and I don't see any real sign of innovation.

One of the chairs is covered in a vinyl imitation leather material. Since it is very hot (I am writing this in June), this chair is not comfortable, and I do not use it. Instead I use a chair covered in a woven cloth material which I find much cooler and more comfortable. So the designer of the first chair failed to solve the problem of making the chair comfortable. The chair I am sitting on is made of tubular steel, and is quite light. So the designer of this chair solved the portability problem. This chair was also inexpensive, so its designer solved the cost problem. So, am I sitting on the perfectly designed chair? I'm afraid not. It's a little hard, so the designer did not get a perfect solution to the comfort problem. But it might suit someone else very well!

SAQ 1.2

The factors fall into broad groups. Each group will be likely to contain many sub-factors. You might have thought of the following.

Functionality. What features should the telephone offer?

Human factors. Usability by a wide range of people. Does it suit the physical sizes of various people (head, hands and fingers)? Is it understandable and clear how to operate it?

Costs. Can it be made at a cost which is acceptable in the market place? Can costs be reduced anywhere?

Materials. What materials are compatible with the expected life-span? What might the buyers want or need? What materials might assist the protection of the internal components? What materials are compatible with mass manufacture?

Image. What form might be attractive to the market? Does it suit the intended context (home or office etc.)?

Manufacture. Is it specified how the phone must be manufactured or is there flexibility to exploit new processes? How many telephones are to be made? Can assembly costs be reduced?

Marketing. Where and how will the telephone be sold (mail order, shops, Internet)?

Does packaging and marketing have an influence on product design?

SAQ 1.3

(a) You might want to produce some images of what the new deodorant container might look like. This would provide general feedback about the visual qualities but if you required feedback on the feel and use of the steel package you would probably have to have full-size models made. Wooden models might be easiest to make but it would then be difficult to convey the cold feel of steel. There might also be the need to model the pressure inside the can, to ensure that the steel wall is thick enough.

(b) Similarly with the car. If one only requires the initial impression of potential buyers then renderings or computer-generated images may be appropriate. Full size, highly realistic clay models are still used in the motor industry to facilitate feedback from potential users and for evaluation by the design team and senior management. If feedback on comfort is required then models using real seats, a dashboard, and a steering wheel might be used. Because such a model lacks the outer skin of the proposed car, it will look very different from the clay appearance-model.

(c) The production of a swim suit is relatively low-cost. Some initial computer-based mapping of colours and textures onto a virtual human form may take place, but it is reasonable to make up a prototype and have it worn by a real person.

(d) The elevated road is clearly too big and too complicated to model at full size. Even if you did make a full-sized model, what would it tell you? The information you actually require is likely to concern costs, material volumes, changes to traffic flow, changes to pollution etc. and these can be more successfully modelled via computer-based statistical techniques or specialist programs for civil engineering. You might also construct partial models to investigate, for example, the stress in a particular road section or support structure. A scale model of the design might be helpful in communicating the proposal to local residents. Increasingly, virtual reality can provide useful models for generating meaningful feedback about architecture and engineering.

SAQ 1.4

(a) None of them has a unique solution.

(b) I will take each case in turn.

  (i) Designing a house for a client. This would involve finding a problem–solution pair – as discussed in Section 1.8. Specification of the 'problem' – the requirements of the client – will give some idea of the solution, but there will be a myriad of solutions in the end.

  (ii) Designing a ball gown for a princess. This probably requires a problem–solution pair. The princess and dress designer would probably work up the design between them making many changes, until they were happy with the result. If the designer were working in isolation, the 'problem' would require greater specification: should the gown be low-cut, have a high hemline etc.

  (iii) Designing a bracket to support a shelf. In this case the requirement is essentially fixed and cannot be negotiated or changed. So no problem–solution pair. There might be different solutions, but only one problem.

(iv) Designing a six-lane road bridge to cross the river Severn. Again the requirement is fixed. There might be some negotiation about the form of the bridge, or even its position, in which case this might mean finding a problem–solution pair.

(v) Designing a railway locomotive. This probably would involve some negotiation about the price and performance, and tests during manufacture might require modifications to the original specification. Also, problems encountered during the construction might requirement modifications to the specification.

SAQ 1.5

(a) Four models of the design process were generated, each adding something to the previous one.

(b) Each of these four models was evaluated. The first three were considered inadequate in various ways. The final model was evaluated as satisfactory.

SAQ 1.6

See Figure 1.65. It would be possible to make the diagram even more detailed. Inside each of the 'generate' boxes there is a generate–evaluate cycle by the designers and the commercial directors of the company, before they decide that the design is ready to go to market, where it will be tested by consumers.

Clearly the spiral diagram in Figure 1.65 gives only gives an approximation to what happened in the evolution of design of plastic kettle designs.

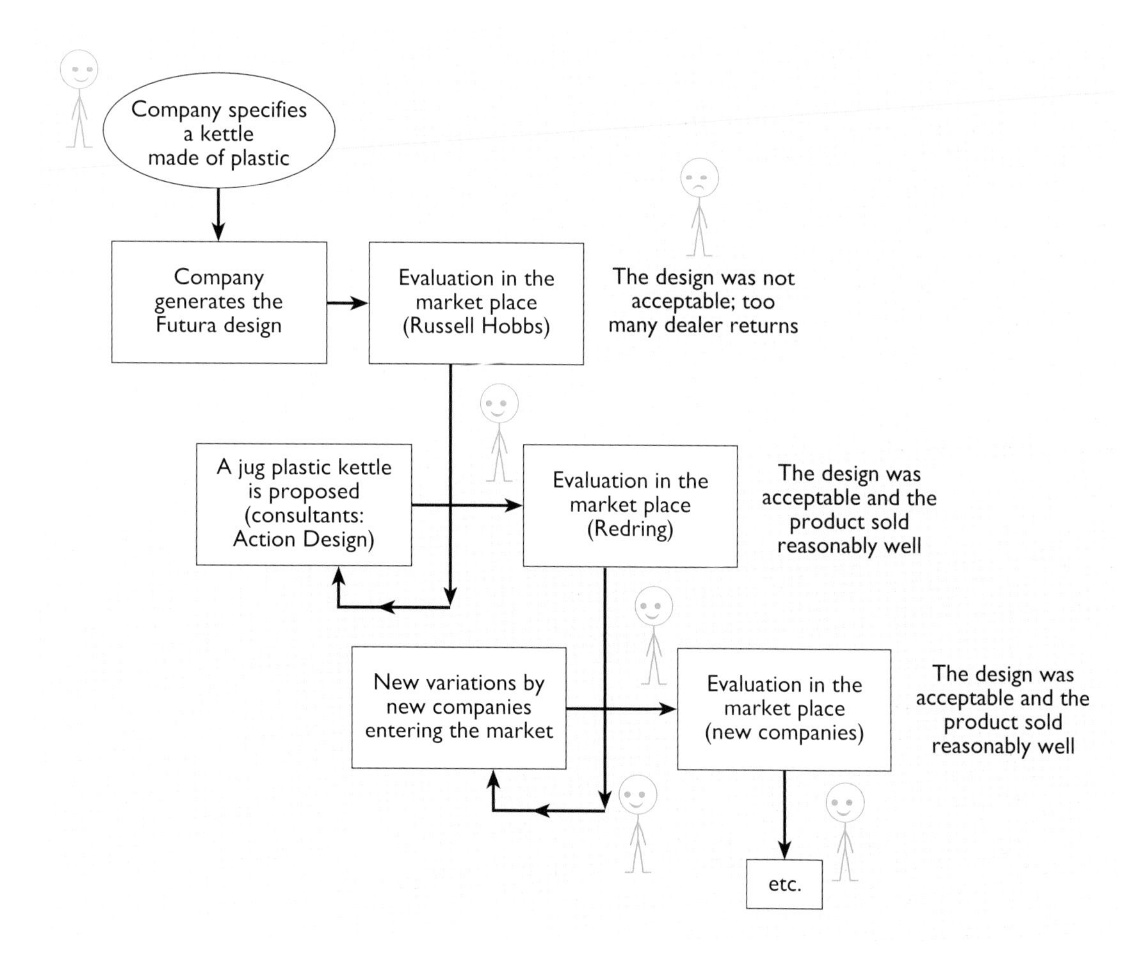

Figure 1.65 Spiral diagram for the development of the plastic kettle

SAQ 1.7

The *trigger* would be the need or desire for a new house (or the desire to make money if you were a developer).

The *product planning* would take place between you and the architect.

You might do some sort of *feasibility study*. For example, a few phone calls might make it clear that you cannot borrow as much money as you had hoped, so that it might not be feasible to find a solution to the original specification (swimming pool, tennis court, five en-suite bedrooms...). Something's got to go.

Then the *design* would be done by the architect, who might develop your favoured scheme into a final design.

This is handed on to the builders to *develop* and *produce* your house.

In this case your house will not be transported anywhere so there's no *distribution*.

Then you'll start to *use* it.

You probably will not have to worry about *disposal*, since the house will probably outlast you!

SAQ 1.8

(a) Well, I've never designed a jet aircraft, but here goes.

The *trigger* is presumably the desire to make money by carrying people long distances at a reasonable cost in reasonable comfort. Perhaps holiday trends or business travellers' routes have changed.

The *product planning* would involve deciding how many passengers, how much they might pay per seat, what distances they want to travel, what comfort they might expect and so on. An outline design specification would have to be produced, e.g. size of aircraft, number of engines. There would probably be some basic visuals produced such as sketches or images on computer.

The *feasibility study* would probably involve a lot of calculations to see if the aircraft was financially viable. Also many calculations to see, for example, which engines have enough power to propel the fully laden aircraft.

The *design* would be very complicated. I have in mind hundreds of designers sitting at computer-aided design (CAD) workstations. Everything has to be designed, right down to the personal light switches although I suppose some items might be already available 'off the shelf'. I begin to think the model is rather weak here, because a huge amount of activity has to go in this box. Lots of models would be used such as CAD models; full-size mock-ups of, say, the cabin; and mathematical models.

The *production* would take place in a hangar, although many components would be produced elsewhere. I would expect to *test* the various components of the aircraft to make sure it worked properly and was safe. The various models would have been thoroughly tested during design so I shouldn't learn anything new at the first test flight!

The *distribution* would include delivering the aircraft, probably by flying it to the customer, and commissioning it. This might involve training the new pilots and maintenance crew.

The *operation* would involve the carrier flying the plane. It would also involve a great deal of maintenance, which I don't see on this diagram.

In one sense *disposal* might refer to the carrier selling on the aircraft to another carrier, rather like selling a car before it becomes expensive to maintain and while it still has high value. Ultimately disposal would probably take place at an aircraft scrap yard.

(b) I think aircraft designers probably do have to go back to previous stages. For example, what happens if the product fails at the design stage (perhaps the models reveal that the required efficiency cannot be achieved) or at the production stage (the components are too expensive)? Then again, it is presumably 'back to the drawing board' and some earlier stage.

It is not uncommon for machines to fail unexpectedly when they are in operation (that is, in ways that the models did not predict). For example, sometimes all aircraft are grounded until some safety device is retrofitted. The redesign of the faulty feature also means going back in the design process. The new knowledge might also lead to better modelling and testing procedures.

(c) I don't feel the BS 7000 diagram has represented my aircraft design process very well. It's disappointing to have 'design' as a box without further elaboration. And the suggestion that one does not 'loop back' is simply not realistic.

SAQ 1.9

Deciding good and bad points is a very personal thing. Table 1.6 gives my lists.

Table 1.6 Comparison of design models

French	*Pahl and Beitz*
Simple, easy to understand at a glance.	Quite complicated, requires studying.
I think this scheme would work quite well for house design.	I think this scheme might work better for house design.
In summary, the procedure is to analyse and state the problem, sketch out some conceptual design, select those that look most promising.	All of this, but in more detail.
I'm not sure what 'embodiment' means for house-building – we can only do it once, so this does not seem to fit our problem.	There is more explicit optimization in this model. That's important because there will be many trade-offs in house design.
Certainly the design has to be detailed into a set of drawings, so that the planning office can pass it, and the builder can build it.	Again the plans and drawings are discussed in more detail.
Does not allow the perceived 'Need' to change through time. Recall the discussion of design as being a process to find a problem–solution pair.	Does not allow the perceived 'Task' to change through time. Recall the discussion of design as being a process to find a problem–solution pair.
'Needs' often have to be changed in house building, because the loose starting specification may have no solution.	Similar remark.

Although I found French's picture simpler and initially easier to understand, I subsequently found the Pahl and Beitz picture to be more satisfactory because of the greater detail it gives. But it depends on the purpose. If I were trying to explain design to a lay person I'd probably prefer to start with French's picture. If you preferred French's picture because of its greater simplicity, that is a perfectly sensible judgement. Better to have a simple picture that you understand, than have a complex picture which you cannot understand.

SAQ 1.10

The long-keel design has more available space within the keel, so it is not necessary to build as large a superstructure on the deck. For equal available space, a fin-and-skeg design requires more superstructure to be built above the level of the deck.

SAQ 1.11

(a) Models assist the definition of the problem and they provide tangible ways of communicating and testing possible solutions.

(b) In the example of sailing boat hulls there are a number of models, in the form of accepted types of hull design, which have proved themselves over many years. The designer doesn't have to start with a blank sheet of paper. Where the designer intends to use conventional materials then the problems are well known and the designer can move on to models which assist the testing of ideas.

However, where new materials or processes are involved (such as the introduction of steel or new polymers for hulls) then mathematical models may need to be used to explore the material properties – especially if they have not been used in hull construction before.

You would probably still find quite a bit of sketch modelling also going on.

Computers are now used extensively to model variant designs based on the basic principles presented in Figure 1.23. They can provide a capability for mathematical modelling (e.g. buoyancy testing, material weight and efficiency in the water) as well as providing images for visual evaluation and marketing.

Testing with scale models is rarely undertaken these days – they may be nice to look at but they are just not accurate enough to provide the information required for design. Sea-testing a prototype hull may also be considered part of the modelling process.

SAQ 1.12

(a) There are many solutions!

(b) Perhaps you might have used the following:

(i) Look at failures and avoid similar layouts of connection.

(ii) Look ahead in your 'mind's eye' to connections in the next module.

(iii) Make a modification to a failure.

SAQ 1.13

(a) Perhaps uncomfortable seats in our car, awkward adjustment and securing of the straps on a bicycle helmet, packaging that is difficult to open, difficult access to buildings and many more.

(b) For the bicycle helmet the adjustment needs to be secure since the helmet must stay in place. Further difficulty may arise from two loose ends requiring two-handed operation. Perhaps fix one end of securing clip to helmet and incorporate adjustment here?

SAQ 1.14

Two ways of changing the stiffness would be:

(a) Use a stiffer material.

(b) Make the tube thicker. This could be done by making the wall of the tube thicker, or by increasing the diameter of the tube. Both methods would increase the stiffness.

SAQ 1.15

You will find that the wooden ruler is stiffer than the plastic ruler. (If you are using a metal ruler, you will find it is not very stiff, but you should notice that it is also much thinner than a wooden or plastic ruler.) From this you can observe that stiffness depends on materials and that wood is stiffer than plastic. You would also find that wood is stiffer than plastic if you were sensitive enough to feel the deflection when pulling the rulers apart. There is another general observation that can be made from this experiment; it is that tension is much better resisted by a structural member than is bending. Try breaking a matchstick by pulling it apart.

SAQ 1.16

The area of a circle is πr^2, where r is the radius (equal to half the diameter) of the rope. So for the 5 mm diameter rope the cross-sectional area is:

$$\pi \times \left(\frac{0.005}{2}\right)^2 \text{m}^2 = 2.0 \times 10^{-5}\ \text{m}^2$$

For the 25 mm diameter rope, the area is:

$$\pi \times \left(\frac{0.025}{2}\right)^2 \text{m}^2 = 5.0 \times 10^{-4}\ \text{m}^2$$

The cross-sectional area of the 25 mm diameter rope is 25 times larger than that of the 5 mm diameter rope. Although the diameter (and so the radius) is only five times larger, because the area depends on the square of the radius, it is larger by $5 \times 5 = 25$ times.

The stress is given by:

$$\sigma = \frac{F}{A}$$

So for the 5 mm diameter rope, the stress is:

$$\sigma = \frac{F}{A} = \frac{500}{2.0 \times 10^{-5}}\ \text{N m}^{-2} = 25 \times 10^{6}\ \text{N m}^{-2}$$

And for the 25 mm diameter rope, the stress is:

$$\sigma = \frac{F}{A} = \frac{500}{5.0 \times 10^{-4}}\ \text{N m}^{-2} = 1.0 \times 10^{6}\ \text{N m}^{-2}$$

Because the area of the 25 mm diameter rope is 25 times that of the 5 mm diameter rope, the stress is correspondingly 25 times smaller. Hence the larger rope can carry a force 25 times greater than the smaller rope before the strength of the rope is exceeded and the rope fails.

SAQ 1.17

(a) For the 10 centimetre bar which is extended by 1 centimetre,

$$\varepsilon = \frac{\Delta l}{l} = \frac{0.01}{0.1} = 0.1$$

As a percentage:

$$\varepsilon = \frac{0.01}{0.1} \times 100\% = 10\%$$

(b) For the 100 centimetre bar which is extended by 1 centimetre,

$$\varepsilon = \frac{0.01}{1} = 0.01$$

As a percentage:

$$\varepsilon = \frac{0.01}{1} \times 100\% = 1\%$$

Note that the *elongation* is the same in both cases, but the *strain* is different because the bars were of different lengths.

SAQ 1.18

From Figure 1.55, at a stress of 40 MN m^{-2} the strain is around 0.06% for the aluminium. The slope E is therefore given by:

$$E = \frac{40 \times 10^{6}}{0.06 \times 0.01}\ \text{N m}^{-2} \approx 70\ \text{GN m}^{-2}$$

So the Young's modulus of aluminium is around 70 GN m^{-2}.

SAQ 1.19

The bottom row of Table 1.7 gives the merit index for each material.

Table 1.7

	Aluminium	*Steel*	*Titanium*
Density ρ /kg m^{-3}	2800	7900	4500
Young's modulus E/GN m^{-2}	70	210	110
Merit index $(E \div \rho)$/MN m kg^{-1}	25	27	24

Note that the units for the merit index are rather peculiar. They are dependent on the criterion which has been selected for the merit index.

SAQ 1.20

The bottom row of Table 1.8 gives the merit index for strength.

Table 1.8

	Aluminium	*Steel*	*Titanium*
Density ρ /kg m^{-3}	2800	7900	4500
Strength σ_f/MN m^{-2}	350	700	300
Merit index $(\sigma_f \div \rho)$/kN m kg^{-1}	125	89	67

SAQ 1.21

Depending on which example you took, you may have come up with any of the following points:

(a) There are many possible developments of the initial concept.

(b) The requirements may be vague; the specification may be poor.

(c) The time and effort to bring the product to market is not known accurately.

(d) Final market demand for the product is uncertain.

SAQ 1.22

(a) Plastic kettle – new technology (materials) in existing context.

(b) Bridges (e.g. the Millennium bridge) where conditions of use or operating conditions are not known completely in advance, leading to failure.

(c) Sony televisions – same technology and function but different look. Folding bicycle has style but this is very much a product of close attention to function.

(d) Stretcher carrier has 'a particular closely defined' context producing a functional design. Elaborate context of soap powder packaging where design has numerous contexts of market, competition and different production requirements.

References

Cross, N., Elliot, D., and Roy, R. (1973) *Man-Made Futures – Readings in Society, Technology and Design*, Hutchinson Educational.

French, M. J. (1985) *Conceptual Designing for Engineers*, Design Council, London.

Hadland, T. and Pinkerton, J. (1996) *It's in the Bag*, Hadland.

Maccoby, M. (1991), 'The innovative mind at work', *IEEE Spectrum*, pp.23–35.

March, L. J. (1984) *The Logic of Design*, in N. Cross (ed.) *Developments in Design Methodology*, Wiley, Chichester.

Pahl, G. and Beitz, W. (1996) *Engineering Design*, 2nd edn., Springer, London.

Ritchie, A. (1999) Private interview with Adrian Demaid of the Open University, 11 March 1999.

Sobel, D. (1995) *Longitude*, Walker Publishing Co. Inc.

Part 2
Design and Innovation in Space

Contents

1 Introduction

In Part 1 of this block, you saw several case studies which were focused on the design of the 'product'. This part of the block uses the example of space exploration and commerce – a dangerous and expensive field – to explore the design process, including its management and development. The constraints, engineering and financial, and the technical complexity of space missions mean that the design process is more complicated than for a relatively simple consumer product.

In this part, I will explore how engineers go about designing satellites and vehicles for space applications. I will describe the design processes and organizational structures that engineers use to meet the demanding requirements of scientists and business plans in the space domain. You will see how the 'vicious' circle of complexity and expense that marked the first thirty years of the space age is giving way to a 'virtuous' circle of 'faster, better, cheaper' space missions in the early years of the twenty-first century.

At the end of this part you will be able to explore some of the challenges facing entrepreneurs who want to make profitable business in Earth orbit. Understanding the funding of development projects is a key part of understanding the design process as it occurs in high-risk industries.

First I will describe why it is that we would want to design anything for space.

2 Why design for space?

> Man must rise above the Earth – to the top of the atmosphere and beyond – for only then will he fully understand the world in which he lives.
>
> Socrates, *circa* 450 BC

As Socrates noted, space provides the ultimate vantage point from which we can observe our planet (and other celestial objects). Space also offers other compelling advantages:

- the opportunity for satellite communications, for broadcasting messages around the world, and other applications such as global positioning;
- an abundant resource of solar energy and extraterrestrial materials;
- a 'zero-gravity' environment for manufacturing new materials and products;
- a vast natural laboratory for physics, chemistry, biology etc.

In this part of the course, we will look at the reasons for going into space, the commercial benefits it offers, and what messages and models of the design process can be extrapolated from projects developed for this unique environment.

2.1 The high ground

On Christmas Eve in 1968, the Apollo 8 spacecraft with its crew of Frank Borman, Jim Lovell and Bill Anders became the first space mission carrying humans to escape Earth's gravity and enter that of another celestial body: the Moon. They were able to look back and see the unique view of the Earth rising over the Moon's horizon.

Closer to home, there is nothing surprising about the fact that, looking from a spacecraft orbiting the Earth, much or all of the globe (depending on the altitude of the orbit) can be seen. Soon after the first satellites were put into orbit, the importance and implications of their potential for observation became clear. The key to the importance of Earth observation from space is the large-scale view that is obtainable from a satellite. Moreover, satellites can be placed in orbits such that they repeatedly cross over every point on the globe; such a satellite can pass over the same spot on the Earth at the same time of day on a cycle of as few as 18 days.

Figure 2.1 Earth observation from space

Nowadays, Earth-orbiting satellites provide us with accurate and up-to-date maps and environmental data for anywhere on the Earth. For many years, pictures from weather satellites have been familiar to television viewers. Moreover, the number of fatalities arising from the annual cyclone season in the Bay of Bengal has been greatly reduced by the much better forecasts that have become available using information from these satellites. They are also an effective tool to study and monitor the global environment: agricultural and geological resources, pollution, desertification, deforestation, fish stocks and climate change to cite a few. Table 2.1 provides a comprehensive list of sectors in which Earth observation satellites play a part.

Table 2.1 Sectors served by Earth observation spacecraft

Agriculture/forestry	Geothermal activity	Sea state
Air pollution	Grazing areas	Snow lines
Archaeology	Harbour effluents	Soil classification
Atmospheric surveying	Hydrographic data	Storm warnings
Conservation surveys	Icebergs	Traffic patterns
Crop census	Irrigation systems	Tree counting
Crop surveys	Livestock census	Urban development
Dam sites	Map making	Vegetation monitoring
Earthquake prediction	Mineral surveys	Water pollution
Fish farming	Nuclear power station sites	Weather monitoring
Fisheries management	Ocean transport routeing	Wildlife surveys
Flood surveys	Oil exploration	
Forest fires	Rainfall monitoring	

Partly as a result of the effects of satellite observations, the term *Earth resources* has come to mean not only crops, forests, water and minerals, but also qualities, features and constituents of the planet of interest to humanity.

There are also military uses of remote-sensing satellites. From the *high ground* of Earth orbit, military forces have been able to keep a very close eye on activities on a global scale. The ability of 'spy satellites' to read the headline of the newspaper you might be looking at in the garden is only a slight exaggeration!

Then there are the scientific instruments that have been put in space to offer a clear view of the Universe, the best known example being the Hubble Space Telescope (Figure 2.2). Observing the universe from the Earth's surface is

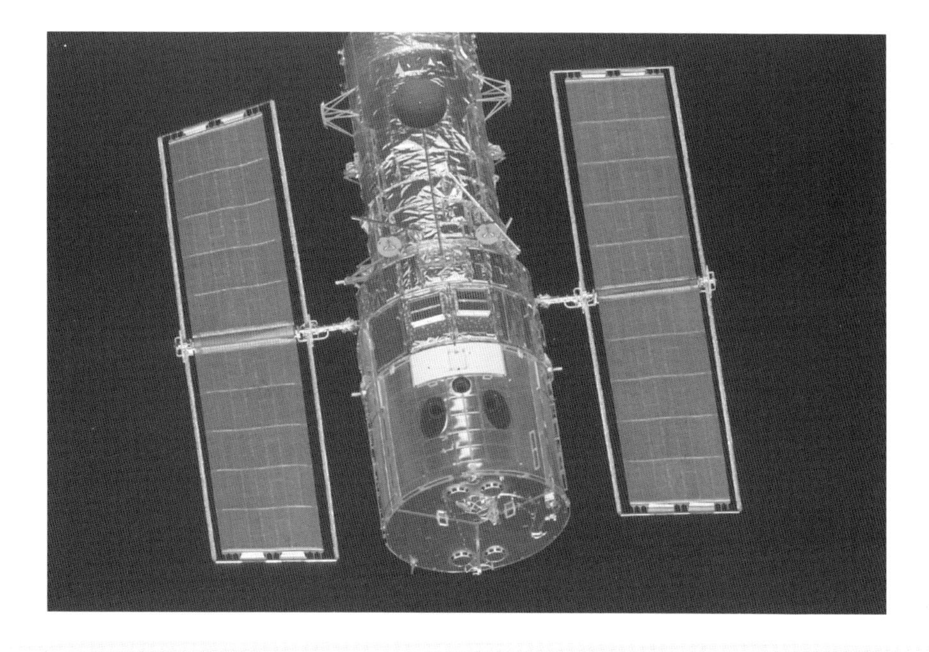

Figure 2.2 The Hubble Space Telescope

hampered by the presence of the atmosphere, which can scatter and absorb light from distant stars. This attenuation is frustrating for astronomers who need to detect and analyse faint objects in order to explore the Universe comprehensively.

The possibilities of Earth orbit attracted the attention of Arthur C. Clarke in 1945; Clarke has since become famous for his science-fiction novels. However, in 1945 he published in *Wireless World*, a magazine for radio enthusiasts, a paper called 'Extraterrestrial Relays'. This paper set out the basis for the global satellite communications network of today.

> One orbit, with a radius of 42 000 km, has a period (the time it takes to go around the Earth) of exactly 24 hours. A body in such an orbit, if its plane coincides with that of the Earth's equator, would revolve with the Earth and would thus be stationary above the same spot...[A satellite] in this orbit could be provided with receiving and transmitting equipment and could act as a repeater to relay transmissions between any two points on the hemisphere beneath. A transmission received from any point on this hemisphere could be broadcast to the whole visible face of the globe.
>
> Clarke (1945)

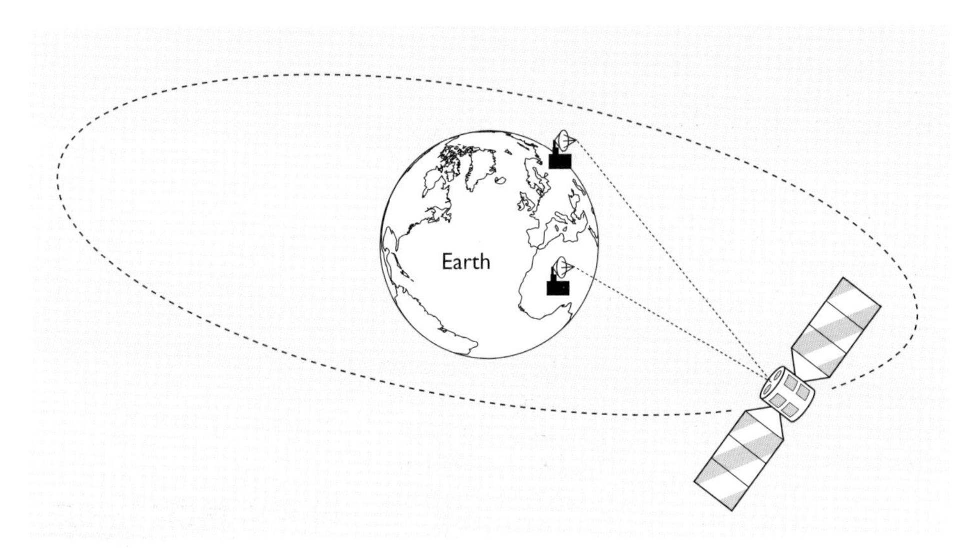

Figure 2.3 Clarke's idea for satellite communication. A signal from the ground is relayed to a distant point via the satellite

The orbit Clarke was thinking of is now called 'geostationary orbit'[1], as the satellite permanently occupies a position above a particular point on the Earth's surface and therefore appears stationary from the Earth.

The first step in turning Clarke's idea into reality was taken in 1960 when the first experimental communications satellite, Echo 1, was put into orbit. Although Echo 1 was really only a large reflective balloon in low Earth orbit, the bouncing of radio signals off it suddenly awoke the world to the potential of communications by satellite.

If several satellites are placed into geostationary orbit, signals can be broadcast to anywhere in the world, by relaying the signal between satellites, as in Figure 2.4. Nowadays, satellites in geostationary orbit provide voice, data, and video links between regions and between continents, and also act as transmitters for broadcasting television and radio over wide catchment areas.

More recently, advances in electronics have aided valuable new applications of communication satellites. Probably the best known is the Global Positioning System (GPS). A large constellation of satellites orbiting at low altitude is used to provide precision location services to small handsets. Even

[1]There is a growing movement to have it called 'Clarke orbit'.

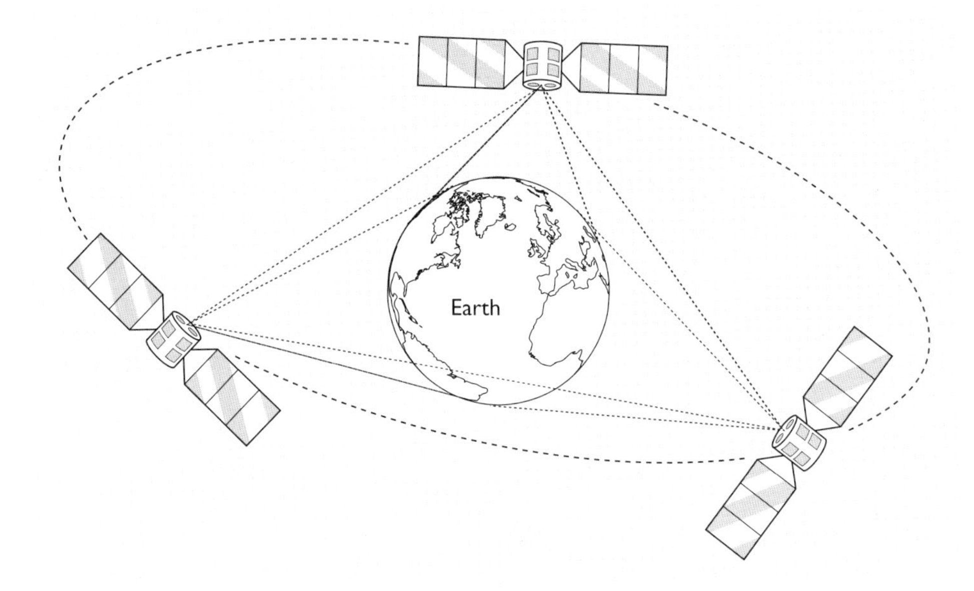

Figure 2.4 Satellite communication using a network of satellites. Now any point on the Earth's surface is accessible

more recently, satellite constellations have been developed to provide mobile telephony and data to handsets from anywhere to anywhere on Earth (except underwater and the deepest caves, to which the radio signals cannot penetrate).

Of course the ultimate goal for many scientific applications is deep space itself. Since the early 1960s robotic exploration spacecraft have been sent to fly-by, orbit and land on other planets and their moons, asteroids and even comets. Mars in particular has been a major target of interest since the early days of the space age. Of all the planets in our Solar System, Mars is the most like Earth. With a thin atmosphere, weather, seasons and a 25-hour day, Mars has a diverse and complex surface, including ice and evidence of water. Although conditions on Mars cannot support life now, a variety of evidence suggests that Mars was warmer, wetter and had a very much denser atmosphere early in its history. Life may have existed. If so, fossil evidence may be found. Mars is where most of the action will be for robotic interplanetary exploration missions in the first decade of the 21st century. We will return to this later in the course.

Looking further afield, the moons of the larger planets, particularly those of Jupiter and Saturn, offer fascinating possibilities. Europa, for example, one of Jupiter's moons, may have an ocean buried beneath its icy crust. Missions to these distant moons are being planned at the time of writing (2000).

Much further away in deep space, a spacecraft launched on 3 March 1972 is still on its way to the stars. On 5 June 2000, Earth's first deep space explorer, Pioneer 10, was so far away that radio waves (or light) from the Earth would take 10.5 hours to reach it.

Figure 2.5 The Pioneer 10 spacecraft

Exercise 2.1

Light travels at a speed of 3×10^8 $\mathrm{ms^{-1}}$ in a vacuum. How far away was Pioneer 10 on 5 June 2000?

The answer to Exercise 2.1 shows that enormous distances can be notched up relatively quickly in space. This is the reason that distances in space are often quoted in 'light years'. A light year is not a unit of time, but a unit of distance. It is the distance travelled by light in a year, and is about 9.5×10^{15} m. Pioneer 10 may be a long way off, but it will still take over 20 000 years to travel the same distance as light can manage in a single year.

▼Design life▲

When designers start work on designing a product, they have to decide what its 'design life' will be – how long the product should be able to function before replacement. Components for cars have to be changed after a certain number of miles; light bulbs often have a life quoted as the approximate number of hours that they will work for.

In the case of a spacecraft, the engineers have to produce designs for on-board power generation and storage, propulsion, and other components that wear out. They also have to think about redundancy in key components to make sure that the spacecraft will meet its objectives throughout the intended design life. For example, the spacecraft might be a satellite that is required to provide a television service for 12 years. Servicing of satellites is difficult or impossible (the cost of servicing may be more than the cost of a new satellite), and once a spacecraft is in space there may be little that can be done to rectify a fault, particularly if the spacecraft is on its way to Mars.

One limitation on design life for spacecraft is the power output from solar panels. These panels degrade in space because of cumulative damage caused by solar radiation. This degradation can be predicted fairly accurately, so designers choose a panel size that is more than enough at the beginning of the mission, and sufficient when the spacecraft has achieved its 'design life'.

Likewise the batteries needed to store the electrical power when the spacecraft is shielded from the Sun (by the Earth, for example), or when it has a temporary high energy requirement, suffer a gradual reduction in performance with repeated charging and discharging. Again, designers choose battery parameters that still give sufficient performance after the spacecraft has been in space for its design life.

Pioneer 10's designers had in mind a ▼Design life▲ of two years when it was launched. Even at the great distance achieved in June 2000, and with near empty batteries the little robotic explorer was still returning valuable data about the extreme edges of the solar system.

2.2 Resources

Space has abundant resources: the solar system has an enormous store of energy and minerals. Currently, Earth-orbiting satellites and interplanetary spacecraft harness only one of these resources: solar energy. But lunar mineral resources, and even those from Mars and asteroids, could one day power a growing space-based economy. Lunar soil, for example, is rich in oxygen and aluminium: both useful if you can get at them.

Although Earth-based mineral resources are currently adequate, looking further into the future we might see that mining other worlds would be a possibility, particularly for elements which are in relatively short supply on Earth.

2.3 Microgravity

On a satellite orbiting the Earth, objects appear to have no weight – not because there is no gravity (gravity is what holds the satellite in its orbit) but because the satellite and its contents are in 'free-fall' towards the Earth.

Imagine an aeroplane travelling very fast at high altitude. If the engines were shut off, the aeroplane would continue to travel forwards, but it would also fall towards the Earth (it would be in free-fall if it were not for the resistance offered by the atmosphere). The path of the aeroplane would be an arc – technically a parabola – and the occupants of the aeroplane would feel weightless because they would be falling towards the Earth at the same rate as each other and the cabin. The faster the aeroplane was travelling when the engines were shut off, the more gradual would be the initial descent towards the Earth, and the further ahead would be the point where the aeroplane reached the ground.

If the aeroplane were travelling fast enough when the engines were shut off, the aeroplane might not reach the surface until it had looped round the Earth, possibly several times. Now imagine the aeroplane travelling faster still, and being outside the Earth's atmosphere so that there would not be any braking effect from air resistance. The aircraft could go right round the Earth and return to its starting point. It would be in orbit. That is, it would be perpetually falling towards the Earth, but not reaching it because the curvature of the Earth would match the curvature of the flight path. Usually there is a state of near weightlessness inside an orbiting spacecraft, a condition described as 'microgravity'.

Many years of experimentation on the space stations of the former Soviet Union, the Space Shuttle and other spacecraft have provided a firm foundation for new products whose manufacture is not possible on Earth because of the effects of gravity. Space manufacturing offers an environment where any detrimental effects of gravity are removed, such as the separation of liquids of different densities during polymer processing. It is also possible to take advantage of the 'cleanliness' of space, with a ready-made vacuum at hand.

You will read about a potential product from space manufacturing in the case study of Section 5.3.

2.4 Spin-offs from space exploration

Quite often, an area of research will lead to applications which were undreamt of by the workers in the field. No one involved in the development of lasers would have foreseen their use in CD players for hi-fi systems, or in sending signals along fibre-optic cables for international communication.

Similarly, space exploration has created technologies that have found their way into markets and applications far removed from their original concept. Optical coatings for the lenses and mirrors of orbiting telescopes have found their way onto sunglasses; cushioning materials used to relieve the acceleration forces (so-called 'g-forces') on astronauts during take-off have found use in orthopaedic mattresses; thermal insulating materials are being used for emergency blankets and heat-insulating gloves. These are a few examples[2], and as more companies become involved in the exploitation of space, the drive to transfer technology to commercial markets will only increase.

2.5 Who pays for space? Who benefits?

During 1998, for the first time, private sector spending on space exceeded that of governments. In the early days, governments were responsible for all finance and management of space projects, but as applications of space matured and commercial space-based services have become viable, commercial space spending has steadily grown in importance.

Governments have always had two roles in space funding – as *interveners* and as *customers*. As interveners, they act to bring about space activities that are perceived to be good for society as a whole, but which wouldn't occur in a free market left to its own devices. It was governments that funded the first experimental communication and remote sensing satellites. By doing this, they opened the way for fully operational systems, funded firstly by international government cooperatives like Intelsat, and then funded by commercial satellite operators.

Of course, scientific missions and military satellite systems have always been funded by governments. The clearest example of a space programme for the 'public good' was the Apollo Moon-landing programme. It produced many commercially successful spin-offs, and generated many new technologies that are applied across many fields of engineering.

[2]An interesting book detailing a range of applications which have spun off from space technology is given in Baker (2000).

Governments can also be customers for space services as well as interveners. A government may, for instance, act as customer on behalf of the people of a country as a whole. Take weather forecasting, for example. Before satellite images were available, if you wanted to host a party in summer, you might have relied on looking at the sky first thing in the morning to assess what the weather might do. Even better, you may have phoned a friend in a part of the country where 'the weather' was coming from. Nowadays, the country clubs together and 'asks' its government to provide sophisticated satellite images and predictions, and to send the bill to the people of the country through the Inland Revenue. In this way, the government acts as a delegated customer on behalf of its country.

At first glance, the roles of intervener and delegated customer may look similar, but the difference between them is, or at least should be, important in the space domain. As intervener, a government gets involved in space for a variety of reasons including domestic ones (for example, supporting a healthy space industry and 'high-tech' jobs) and for reasons of foreign policy (for example, in order to observe international trouble spots). An intervener's decisions aren't based on supply and demand. An intervener is interested in balancing political gains against project costs to determine the priorities of competing initiatives.

As a delegated customer, government takes a different view. If society wanted even better weather forecasts, government might consider investing in a new generation of satellites. But the Treasury would want to see savings somewhere else, and value for money, so the satellite system would have to prove its worth.

This role of government as delegated customer produces some of the most exciting prospects for space initiatives. Many potential applications of space make economic sense, even if they're not necessarily commercial. One of the biggest and most exciting of these is monitoring of the Earth's environment. As public pressure on governments to understand and manage environmental change grows, there will be good opportunities for promoters of space-based observing systems to provide the global data required. But such projects will need to satisfy the hard-nosed economic analysis of those government agencies involved in environmental modelling.

As with many other aspects of engineering endeavour, the private sector is likely to have an increasing part to play in the development of space over the coming years. This is explored further in the audio band associated with the following Activity.

Activity 2.1 Commercializing space

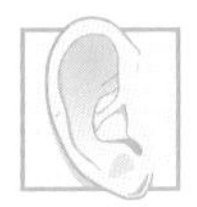

Much of the initial drive for space exploration was tied up with the political aims of the Cold War, with the USA and the USSR vying for technical superiority in space. Space missions now require clear objectives, offering excellent scientific data or the possibility for commercial profit.

The audio band associated with this block was originally broadcast in September 2000 as an *In Business* programme on Radio 4. It investigates the commercial opportunities at the time of broadcast for the short-term commercial exploitation of space. Listen to this band when you have an opportunity, then answer SAQ 2.1.

SAQ 2.1 (Learning outcome 2.1)

Look back over this section, including the audio band, and try to identify the following.

(a) An area of space exploration that is likely to come to fruition in the next 10 years.

(b) An area of space exploration that may come to fruition some time in the next 50 years.

(c) An area of space exploitation (or commercialization) that is likely to come to fruition in the next 10 years.

(d) An area of space exploitation (or commercialization) that may come to fruition some time in the next 50 years.

3 Satellites, spacecraft and systems

3.1 Satellites

The word *satellite* comes from the Latin *satelles*, meaning an attendant or servant to a powerful master or lord. In an extraterrestrial context, the term was first applied to the moons of the planets that were observed after the invention of the optical telescope. These objects seemed to run quickly about the planets, which were themselves named after ancient gods, so satellites seemed an appropriate name.

With the launch of Sputnik in October 1957, the term 'artificial satellite' came to be used for the objects which are placed into orbit around the Earth, and this is an entirely appropriate name for them. These satellites are indeed servants, placed into orbit around the Earth or other planets, to provide a wide variety of important services and information, as we saw in Section 2.

Simply put, a satellite is an object that orbits around a planet, moon, or some other celestial object.

3.2 Spacecraft

The word *spacecraft* tends to conjure up images of sleek flying saucers, or impressive creations from television series. In reality, spacecraft tend to be squat and ungainly rather than sleek and streamlined. The reasons are purely practical – spacecraft are built to perform particular tasks in an efficient, cost-effective manner. In the vacuum of space, there's no need to be streamlined. The term spacecraft includes rather mundane objects such as weather satellites. However, 'spacecraft' is in fact just a more general term than satellite, and means any object which is launched into space for a particular task. The word 'satellite' should be used only when the object is in orbit around the Earth or another planetary body[3].

A spacecraft is often thought of as being divided into two parts – the *payload* and the *spacecraft bus*. The payload is that part of the spacecraft which actually performs the mission function. For example, a spacecraft on a mission to monitor the Earth's ozone layer would have a payload consisting of several instruments designed to measure different aspects of the upper atmosphere.

The spacecraft bus provides all the 'housekeeping' functions necessary to allow the payload to do its work. It provides electrical power, maintains the right temperature, processes information, communicates with Earth (and possibly other spacecraft), controls the spacecraft's orientation, and generally holds everything together.

The spacecraft as a whole has to be highly reliable, lightweight and power-efficient and has traditionally been a costly, often unique, piece of engineering.

3.2.1 Reliability

The high-reliability requirement arises because spacecraft cannot easily be serviced or repaired once launched. The few spacecraft that have been visited by astronauts for servicing or returned to Earth for refurbishment have had a

[3]Purists may point out that if I avoid calling a spacecraft on a trajectory between two planets a satellite, I will have forgotten that the spacecraft is in fact still a 'satellite' of the Sun!

high media profile, notably the case of the Hubble Space Telescope, which was launched initially with a defective lens.

3.2.2 Lightweight

The need for lightweight designs is a result of the huge cost of launching payloads into space. The launch vehicles and their operation cost millions to hundreds of millions of dollars per flight, whether the launch vehicle is reused (like the Space Shuttle) or is expendable, and whether the payload is large or small. This cost for the launch is in addition to the usually very expensive spacecraft that the launch vehicle is carrying. Launch costs decreased rapidly during the first decade of orbital space operations (1957–1967), primarily because of the rapid growth in the size of rocket boosters and a resultant 'economy of scale'. After this initial period, however, launch costs (in the West, at least) flattened out and have remained fairly constant despite the introduction of many new vehicles. There are many launch systems operational at the time of writing (2000), and they cover a wide spectrum of performance, cost, and design and operational philosophies. Table 2.2 gives a summary for the costs of putting a satellite into ▼Low Earth Orbit▲ using these launchers.

The information in Table 2.2 can show some interesting trends in the cost of putting satellites into orbit.

Figure 2.6 shows the cost of putting 1 kg of payload into orbit (vertical axis), in relation to the payload capacity of the launcher (horizontal axis), for most of the launchers in Table 2.2. (Note that 'the cost of putting 1 kg of payload into orbit' does not mean 'the cost of launching a 1 kg satellite'; the calculation of cost per kilogram assumes that the full payload capacity of the launcher is used.)

I have shown two graphs in Figure 2.6. You should recall from Block 1 that it is often useful to use a *logarithmic* axis on a graph when the data spans several powers of ten. You should see that using such an axis makes the data easier to read for the payload capacity.

Table 2.2 Launch costs for a range of vehicles

Vehicle name	*Lift-off mass of launcher/kg*	*Typical cost range per launch/$m*	*Payload capacity to low-earth orbit/kg*	*Cost per kg to LEO/$ kg^{-1}*
Pegasus	19 050	13–15	454	30 800
LLV-1	66 225	15–17	794	20 200
Taurus	81 650	18–20	1451	13 100
Titan II	155 000	35–40	1905	19 700
Vostok SL-3	290 000	20–30	4717	5300
Delta II 7920	218 300	45–50	5035	9400
Atlas IIA	187 700	80–90	6760	12 600
Ariane 44LP	420 000	90–100	8300	11 400
Long arch 2E	464 000	40–50	9210	4900
H-2	264 000	150–200	10 433	16 800
Titan IV	862 000	230–325	17 700	15 700
Proton SL-13	703 000	35–70	20 000	2600
Space Shuttle	2 040 000	350–547	23 500	19 100

▼Low Earth Orbit▲

Most satellites are in what is called 'Low Earth Orbit' (abbreviated as LEO). This means that they orbit the Earth at an altitude below 3000 km. Medium Earth Orbit (MEO) is between 3000 and 30 000 km altitude and some GPS (global positioning satellites) use this altitude range. Communications satellites in geostationary orbit (GEO) typically orbit at an altitude somewhere between 30 000 and 40 000 km.

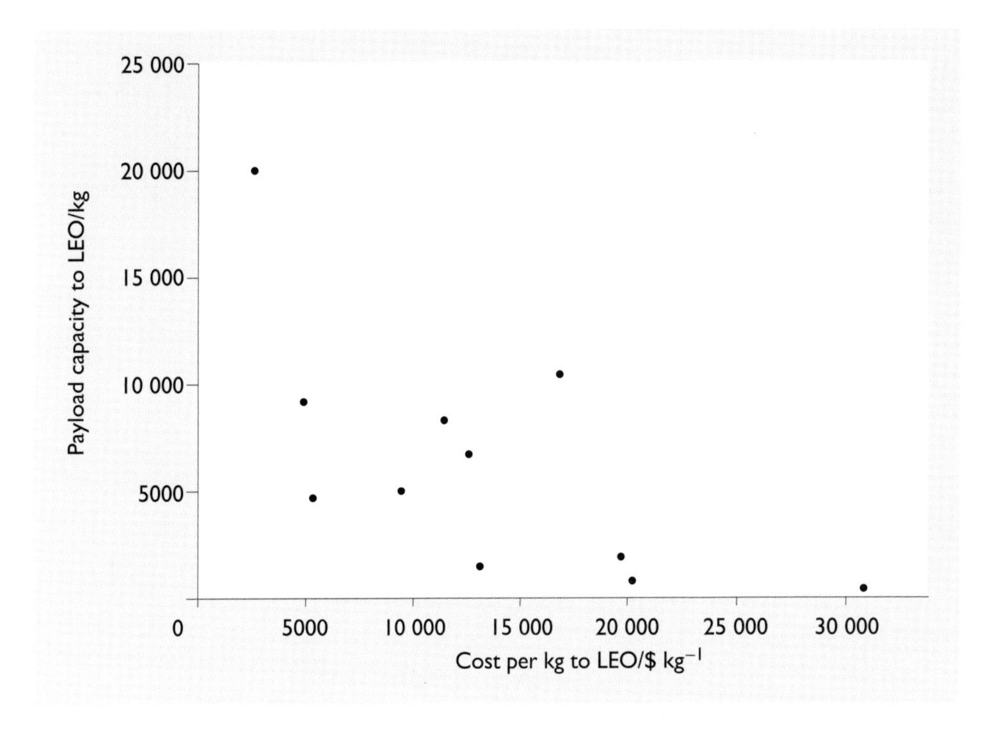

(a)

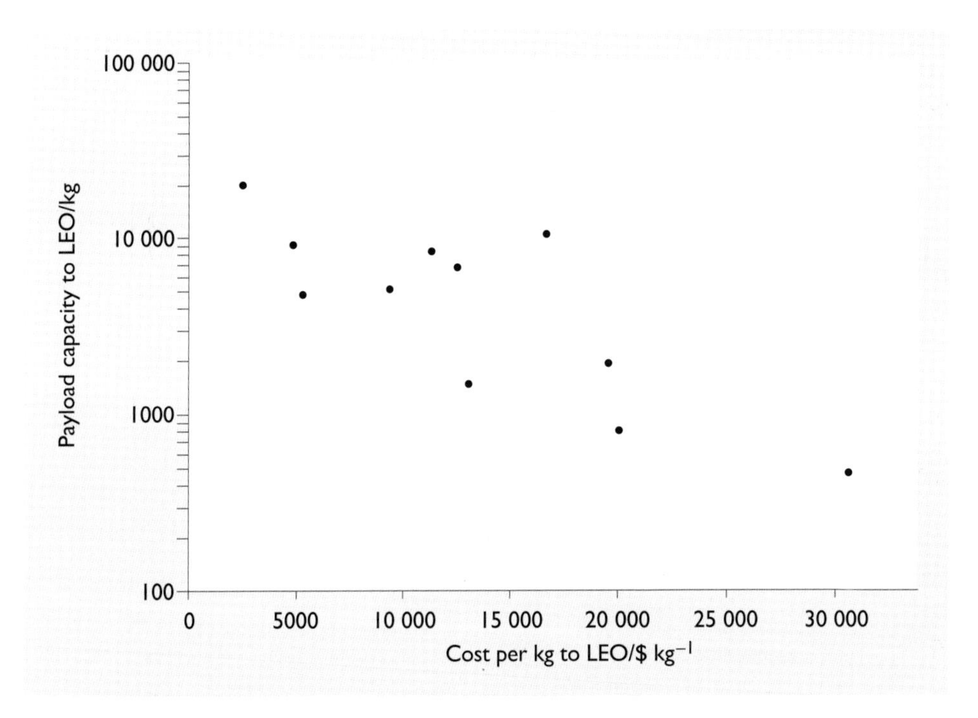

(b)

Figure 2.6 Cost of launching 1 kg of payload to Low Earth Orbit in relation to payload capacity for a range of launch vehicles (a) 'Linear' scale (b) 'Logarithmic' scale

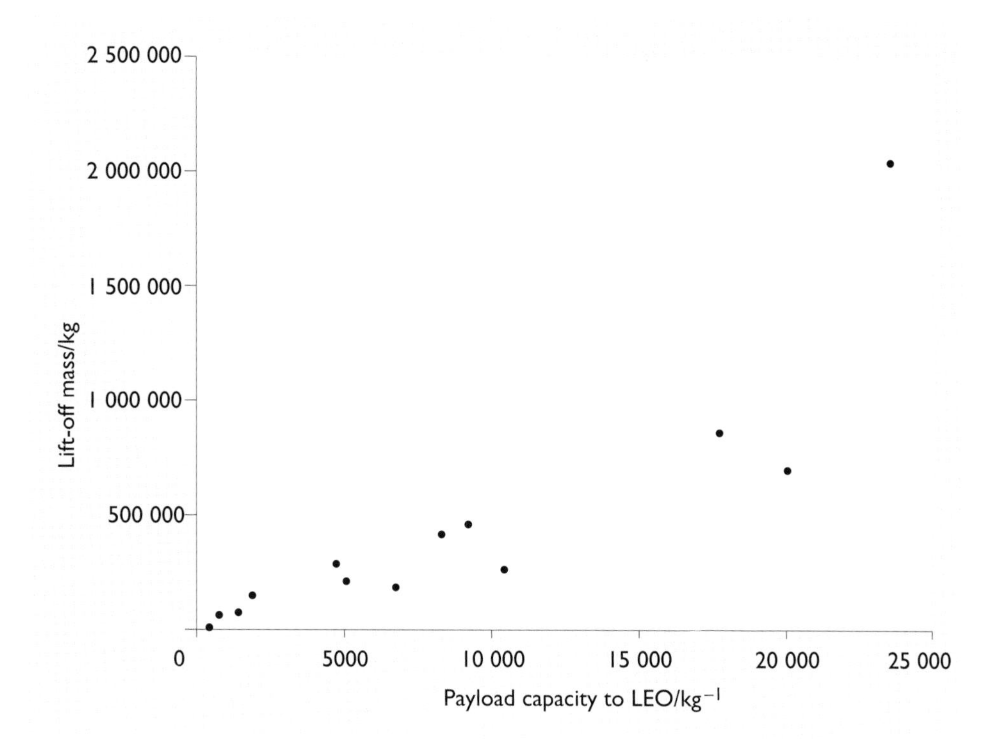

Figure 2.7 Payload capacity in relation to lift-off mass for the launchers in Table 2.2

Figure 2.7 shows the data of payload capacity in relation to the lift-off mass of the launcher.

SAQ 2.2 (Learning outcomes 2.2 and 2.3)

(a) Looking at Figures 2.6 and 2.7, what conclusion can you draw about the cheapest way to launch a satellite into orbit, in terms of selecting a launch vehicle?

(b) I omitted the data for the Space Shuttle and the Titan IV in plotting Figure 2.6. These launchers were devised for a range of missions, involving military payloads as well as commercial and scientific. The Shuttle can also be reused after each launch. Add the data for these launchers to Figure 2.6 (you will find it easier to add them to Figure 2.6(a)). Does that affect the trend that you identified in part (a) of this question?

(c) Why might this difference in launch cost be acceptable for these launchers?

3.2.3 Power efficiency

Spacecraft need to be power-efficient because they nearly always rely on solar power collected by arrays of solar cells mounted on the spacecraft body or on deployable panels. Such cells do not produce enormous quantities of power, and the larger the area of solar panel needed, the heavier the spacecraft will be. So reducing the power requirement for the craft will allow for a smaller solar panel, meaning less weight and a cheaper launch.

3.3 Systems

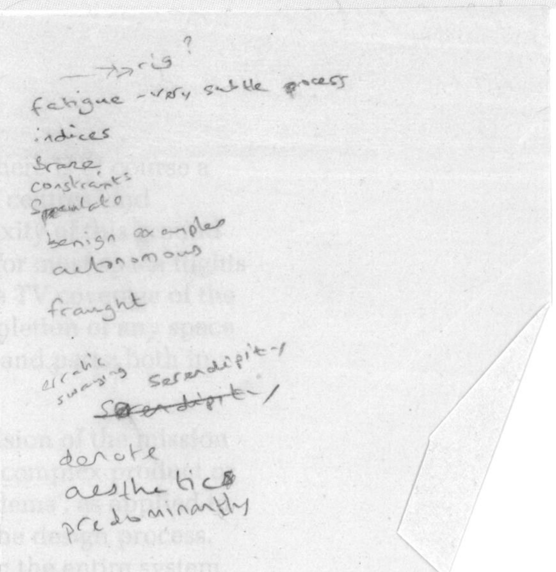

In addition to the spacecraft and their launch vehicles there is of course a 'ground segment' of tracking stations, spacecraft control centres and communications networks. The importance and complexity of this ground segment tends to be somewhat underestimated, even if for manned flights its existence and role are widely publicized through live TV coverage of the important phases of such missions. The successful completion of any space mission requires the close cooperation of all the people and assets both in space and on the ground.

From a mission designer's point of view any simple division of the mission into discrete parts is not good enough. Designers of any complex product or service need to adopt *systems thinking*. The field of 'systems' (or systems engineering, is important for a better understanding of the design process. Thinking in terms of individual components, rather than the entire system, can often be a drawback. The concept of 'systems' is sufficiently important for us to consider it further at this point.

3.3.1 What is a system?

Perhaps the simplest definition of a system is that it is a set of interconnected components. However, when using the term in technology or engineering it is useful to expand on that simple definition. So, we often use the following three-part expansion.

1. A system is an assembly of components, connected together in an organized way, that does something.
2. Each component is affected by being in the system and the behaviour of the system is changed if a component is removed.
3. The choice as to what are regarded as components of the system has been made by someone with a particular interest in mind.

Often the components of a system are complex enough to be called 'subsystems' and can have components or subsystems of their own, maybe down to several lower levels.

When we are engaged in designing something, it can be very valuable to conceptualize that thing, or the situation in which it is to be used, as a system. This is called taking a *systems approach*.

3.3.2 Why is a systems approach important?

At the time of writing (November 2000) the railways in the UK were in a state of chaos. The immediate reason for this was the response of the railway companies to public and Government pressures following two fatal accidents during the course of the year, in which many people had been killed. The second accident was caused by a broken rail, and it was believed to be likely that similar breaks could occur in a number of places throughout the network. To protect lives, speed limits were imposed at these sites and a major programme of rail repairs was instigated to guard against possible further accidents.

In itself, this seems completely rational, but was it? The railways are only part, and a relatively small part, of the UK's *transport system*. During the course of a typical year, there are over 3000 fatal road accidents in the UK, which is about a hundred times as many people as were killed in the rail accidents mentioned above. It is likely that the railway chaos caused by the speed restrictions and repair works encouraged many more people to switch to making their journeys by car.

Some commentators have suggested that the resulting additional road traffic caused extra deaths and injuries; it is possible that the number of such deaths and injuries could be higher than would have been likely to occur if the speed restrictions and other work on the railways had been much less hasty. If Government and the public wished to ensure a safe *railway system*, then the speed restrictions and emergency repair programme was probably the right decision. But if the intention was to ensure a safe *transport system* (which would include road transport), different actions altogether might have been appropriate. It all hinges on choosing an appropriate *boundary* for the system under consideration: the *railway* or the *transport* system.

Let's look in a bit more detail at the three-part expansion of the definition of a system on the previous page. First, we say that the parts of a system are connected together in an organized way, and the assembly does something. So, we would be unlikely to say that a heap of cogs on a bench was a system, since it is not organized and doesn't do anything much apart from gather dust; but those same components connected together in a gearbox could form a 'power transmission system'. The second point is probably also fairly uncontentious, in that assembling the cogs into a gearbox affects the way each cog can move (its *behaviour*) and equally, the train of cogs cannot function properly as a transmission system if one cog is removed.

The third point is perhaps more difficult to understand and accept. When we speak of 'the railway system', or the 'Open University system' we often assume that the term defines some unambiguous set of components, within a universally agreed boundary. But a system, in the technical sense we are using it here, refers not to the components themselves, but to a particular *ordering* or classification of them by someone. The choice of what is included in the list of items, and the way that they are seen to relate to one another, will vary depending on the reason for looking at them as a system, and on the individual who is doing the ordering and classification. So, what I would include in my version of the Open University system (colleagues, the building in which I work, the library and computer network, the warehouse etc.) would be very different from what you, as a student, would probably include. And equally importantly, you and I would have different views of what we want our Open University system to do. I want it to provide me with a job, and a place to work in which I can produce what I regard as useful educational materials. You probably see it as a system to enable you to obtain a different set of outcomes, associated with your learning and possibly your career. So, like beauty, 'system' is in the eye of the beholder. You should recognize in this regard similar ideas to those considered in the earlier material on the design process, where part of the skill lies in choosing what it is (what system) that needs to be designed.

3.3.3 Reductionism, holism and multiple causality

In connection with ideas of systems thinking, you will often come across the related ideas of *reductionism* and *holism*. *Reductionism* is often characterized as being the classic underpinning of science, and is contrasted with *holism*, of which a systems approach is one example.

When using reductionist methods, we take a complex situation or structure, and investigate the behaviour of just one part of it, trying to keep all the other parts unchanged. So, we might take a beam out of a structure, apply a load to it at the point where we believe it would be loaded in the structure, and observe how it behaves as we did in describing the behaviour of different materials for bicycle frames in the first part of this block. We would hope that we had eliminated all the complications that might be introduced by other parts of the structure. This would enable us to obtain a very accurate measure of the response of the beam to the applied load, and would enable us to predict how it would behave as part of the structure, if the load within the structure was actually applied in the way that we assumed. In essence, this is the way that much of science has progressed, although during the twentieth century there were numerous attempts to move science towards a more *holistic* method.

The fundamental idea of holism is the premise that it is impossible, or at least, undesirable, to consider the behaviour of one item in isolation from those other items which interact with it in the original whole. (Hence, 'holism', meaning 'concerned with the whole'.) In the case of the structure, the position at which the load is exerted on the beam may vary as the whole structure is loaded, as a result of deformations in the structure remote from the beam. We would not be aware of this from our studies of the beam's response to loading in isolation. To achieve a satisfactory analysis, we need to consider the structure more holistically. At this level, the idea is plausible and, potentially, achievable with reasonable knowledge of the geometry of the structure, mechanics and the properties of materials. What is more difficult to achieve is the sort of holism that sometimes seems to be advocated by its devotees, which appears to require consideration not just of the structure, but of the weather, the maintenance schedule for the structure, the training of the operators, the way they are paid, their family relationships and so on, apparently *ad infinitum*! Of course, the behaviour of our original beam *could* conceivably be affected by all these things. But in systems thinking we choose to put a boundary on the factors and their interactions, and don't try to consider everything. What we include in our conception of the system is that set of components which is appropriate for what we are trying to achieve.

Another way of looking at the distinction between reductionism and holism is to consider what form of explanation we seek for the *cause* of some event. A reductionist explanation would try to find a single cause–effect relationship, such as we might represent as in Figure 2.8.

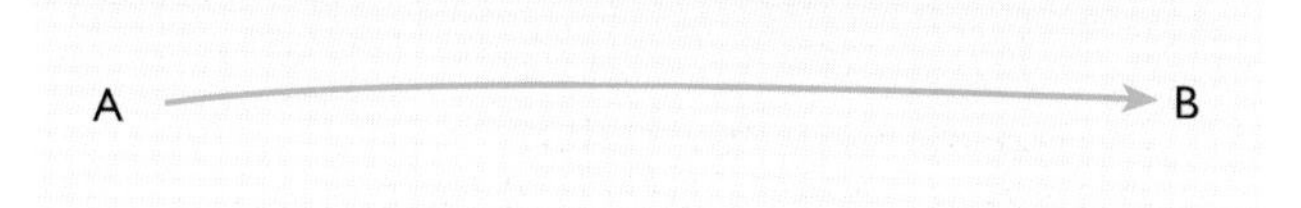

Figure 2.8 A single cause–effect relationship

That is, some factor A can vary, and the behaviour of the item B is caused or affected by, and only by, the variation in A. We illustrate that with the arrow connecting A and B. The arrow means: 'Whatever is at the tail-end of the arrow causes, or at least affects, what is at the point of the arrow.'

In reality it is much more often the case that there are other factors, C, D, etc., that can also affect the behaviour of B; and furthermore, changes in B can cause changes in, say E, that can itself cause changes in C. In such a situation, the behaviour of B has *multiple causes*, as represented in Figure 2.9, which we call a *multiple-cause diagram*. Diagrams like Figure 2.9 are just one of a range of diagrammatic techniques that are useful in representing or thinking about systems.

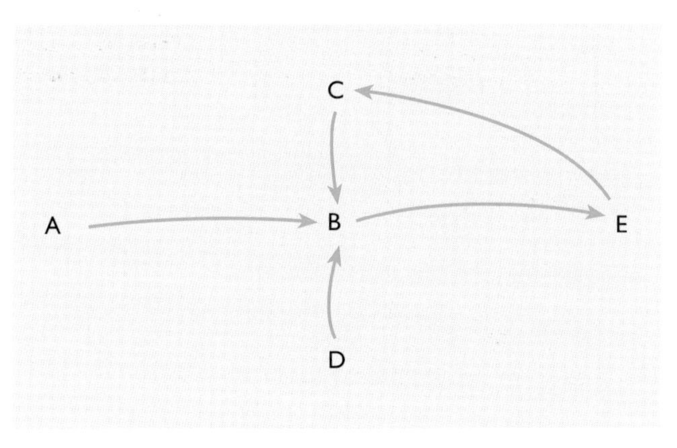

Figure 2.9 A multiple-cause diagram of a more complex network of causes and effects

3.3.4 Systems diagrams

The inherent complexity of systems often means that it is difficult to describe them, and almost impossible to analyse them, using just sequential words on a page. Some mathematical representations can be helpful in particular situations, but a lot can be achieved using diagrams, like the multiple-cause diagram above. From the range of possible diagram types, I want briefly to consider just two. The multiple-cause diagram above is sometimes also referred to as a *causal-loop diagram*, because it includes loops such as C → B → E → C. The items A, B, etc. in such a diagram can be physical entities, measures of the properties of physical entities or even more abstract things such as company morale. Low workforce morale could impair the quality of a product, something which no analysis of materials or structures will tell you. The defining characteristic of all the chosen factors is that it is plausible to consider that each one can cause change in or be affected by, another.

Let's try and use the convention to consider a real example. In Part 1 of this block, we looked at the jug kettle, and at the way that sales of this changed. In the first instance, there was a design idea, influenced by the availability of plastics of particular properties, and methods of producing plastic mouldings. Jug-kettle sales could not occur without these, so we can recognize three possible items to include in our diagram, in Figure 2.10.

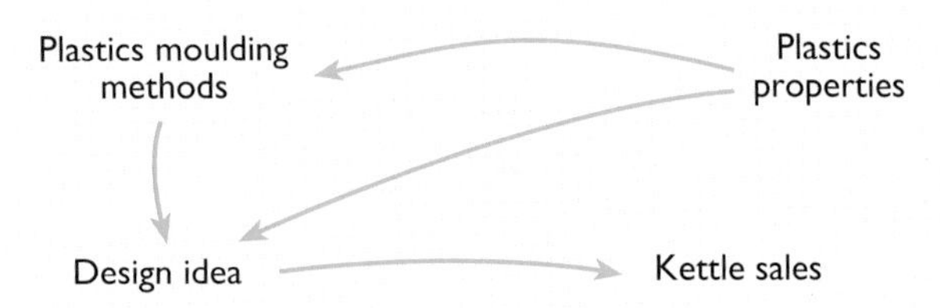

Figure 2.10 A partial multiple-cause diagram of the factors influencing sales of the jug kettle

However, the existence of plastic jug kettles is not sufficient to ensure that sales arise. There is a need for the kettles actually to be manufactured, and for there to be potentially a public interest or demand for such kettles.

SAQ 2.3 (Learning outcomes 2.4 and 2.7)

Modify Figure 2.10 to include these additional factors to show their influence on jug-kettle sales.

From the earlier material, you should recall that sales of the first model of jug kettle were relatively poor, but were enough to create some additional public interest. This can be shown in Figure 2.11 by the additional arrow from jug kettle sales back to public interest.

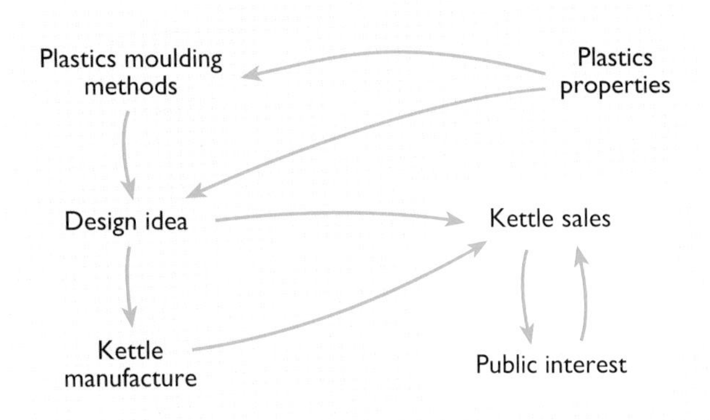

Figure 2.11 The feedback loop from kettle sales to public interest

So, as kettles were sold, more people became aware of them, and so more might be sold. This is an example of a *feedback loop*, which can have dramatic effects on the way a system behaves. In this case, the loop exhibits *positive feedback*, in that sales themselves generate new sales. Although such a situation is desirable in the short term, this sort of growth phenomenon can also give rise to undesirable effects in the longer term, where runaway growth of some sort occurs. The commonest examples occur in biological systems, where pest populations can increase explosively, but there are also engineering examples.

The alternative situation is where negative feedback occurs, so that as one factor increases, the resulting increase in the second factor feeds back to reduce the first, tending to stabilize the situation. This combination of negative and positive feedback is common in most real situations, and can have a major effect on what happens. Often one of the most powerful insights from using a multiple-cause diagram is the recognition of these loops in a particular system.

SAQ 2.4 (Learning outcomes 2.6)

A transport safety example which is often quoted is the use of pinch points or chicanes to reduce traffic speed. These are designed as purely physical barriers, considering the system as the road plus the car. However, there is some evidence that where these barriers have been installed, traffic may actually speed up between the barriers, as drivers race to beat oncoming vehicles to the pinch points.

Draw a multiple-cause diagram to show the relationship of the factors affecting traffic speed on a stretch of road with traffic calming measures in place.

Hint: The appropriate system to consider here should clearly take account of driver behaviour, as well as the simple physical relationship between the car and the road topology.

The second type of diagram which I want to introduce is called a *systems map*, where components of a system, or subsystems, are represented by enclosed shapes. The relationships between the components/subsystems are represented by their spatial relationships. So, for example, a systems map of the UK transport system might have three subsystems: a railway subsystem, a road-transport subsystem and an air-transport subsystem. This is shown in Figure 2.12.

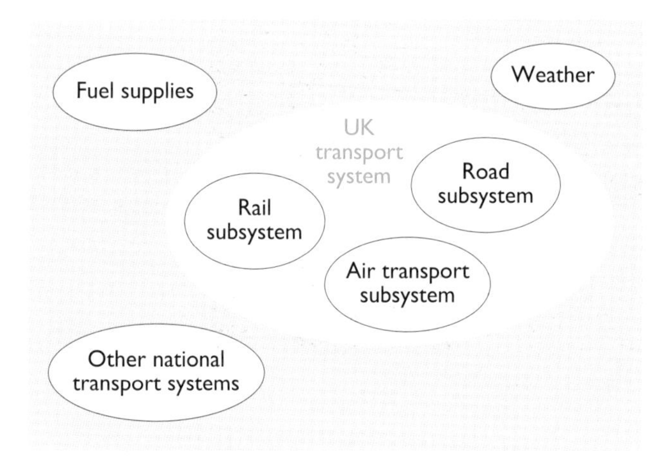

Figure 2.12 UK transport system

Note that I have drawn a boundary around the UK transport system, but that outside that boundary I have indicated that there are other systems. These are those things in the *environment* of our chosen system, which will affect the UK transport system, but are not part of it. So, Figure 2.12 reminds us that it is important to consider the effects of other countries' transport systems on the UK transport system, and also the effects of weather and of fuel supplies. Each of the subsystems in Figure 2.12 is itself made up of further subsystems. For example, the railway system might be considered to have a track subsystem, a rolling stock subsystem, a passenger information system and a ticketing system.

SAQ 2.5 (Learning outcomes 2.7)

Draw a systems map of the railway subsystem from Figure 2.12.

A systems map is obviously a very simple technique, but can be very useful as a way of representing the structure of a system, or in comparing different conceptions of what may at first appear to be a single set of items. We can apply this technique to the activities associated with the space programme to illustrate the general point that, from different perspectives, the system that is perceived within a situation can vary markedly.

3.3.5 Systems view of the International Space Station

The International Space Station (ISS) was first conceived in 1984 by NASA. This huge project to set up a permanent manned presence in Earth orbit has gone through numerous redesigns and internationalization efforts. The first modules of ISS were placed in orbit in 1999. Its eventual configuration will look like that in Figure 2.13.

Figure 2.13 International Space Station (artist's impression)

The individual modules are designed and built by consortia of specialist companies, under the guidance of the international space agencies, including NASA, ESA (the European Space Agency), and Russian, Canadian, Brazilian and Japanese space agencies. Many of the modules have been launched into orbit as the payload of the Space Shuttle, as this vehicle offers the possibility of using astronauts to assemble the modules in orbit. Once the whole station has been assembled, it will be staffed by up to seven astronauts, who will join the station via any of the launch vehicles of the international partners: the Shuttle, the Russian Soyuz craft, etc. Their life support will be provided partly by on-board activities powered by the solar panels shown in Figure 2.13, and partly by supplies sent up from the Earth. These supplies will include materials for replenishing and renewing the Station air supply, water, and pre-prepared food. The component modules of the station include a control module, power modules, various pressurized and unpressurized (i.e. open to vacuum) research modules, a 'service module' to provide life support and living quarters, and a robotic manipulator arm.

SAQ 2.6 (Learning outcomes 2.4, 2.5 and 2.7)

Draw a systems map of the International Space Station from the perspective of a company wishing to design one of the service components of the ISS. Remember to include those items which are important in the environment of the ISS.

Repeat the exercise from the perspective of a company considering investing in an experimental manufacturing process on the ISS. Can you see any potential mismatches between the systems as perceived by the two organizations?

4 Designing space systems: cycles within cycles

What do designers of space systems do and how do they do it?

First we should recognize that designing for space is rather complex. There are many factors to consider, and it is difficult for designers to establish the relative importance of these factors and to generate proposals which seem to offer suitable compromises.

A space project covers a wider range of technology than any other sector of project work undertaken by humanity: from chemical rocket fuel to solar energy to advanced electronics. Its multidisciplinary nature makes it demanding and challenging for any organization concerned with it in any way.

Space missions range widely from communications to planetary exploration to proposals for space manufacturing to burial in space. No single process can fully cover all contingencies, but the design method outlined in Table 2.3 summarizes the approach which has evolved over the first forty years of the space age. Down the left of the table are the broad categories of activity. Down the right of the table, the activities A to J expand on the broad activities on the left.

Table 2.3 The space mission design and analysis process

Define the objectives	A Define broad objectives and constraints
	B Estimate quantitative mission needs and requirements
Characterize the mission	C Define alternative mission concepts
	D Identify the key features of the design
	E Characterize mission concepts
Evaluate the mission	F Identify key mission requirements
	G Evaluate mission success markers
	H Define baseline mission
Define requirements	I Define system requirements
	J Allocate requirements to system elements

As stated in Part 1 of this block, the design process is iterative, gradually refining both the requirements and methods of achieving them. Thus the broad process defined in Table 2.3 is repeated many times for each mission. The first few iterations may take a few days or weeks but the later ones will take months if not years.

Successive iterations through Table 2.3 will usually lead to a more detailed, better-defined space mission design. But designers must return regularly to the broad mission objectives and search for better ways to achieve them. Solutions may change as a result of evolving technology, a new understanding of the problem, budget allocations, politics, or simply fresh ideas and approaches as more individuals become involved in the process.

Finally, designers must document the results of this iterative process. If they wish to go back and re-examine decisions as new data becomes available they must clearly understand and convey to others the reasons for each decision. Thus documentation is needed for decisions based on detailed technical analysis, ease of assessment, or even political considerations.

4.1 Top-down or bottom-up?

It would be a mistake to think that complete designs are produced for entire spacecraft and only then are tests applied. Rather, designers break down the problem hierarchically into several smaller problems, often working in turn on the separate components of a mission, generating alternative designs for these relatively discrete parts and subjecting them to separate tests.

The design of a spacecraft is not the work of one person but of a team. The subdivision of the whole problem into smaller parts will correspond broadly to the division of labour among members of the team. The completed component designs are assembled and further tests of performance applied to the assembly as it is built up. The results of these tests may show that it is necessary to go back and redesign some of the components.

Although the parts into which the overall design problem is broken down might often correspond to distinct components, in other cases the hierarchical breakdown of the problem might not be of this character: more than one design engineer may work on one subsystem, and an individual designer may be involved with several subsystem projects.

The design process goes on then at several levels: on the level of the spacecraft as a whole; on the level of separate subsystems or components, and also on intermediate levels between these. Always the designers will move back and forth between consideration of the parts and consideration of the whole. It is nevertheless possible to distinguish different *styles* of designing.

For example, there may be a process which works predominantly *bottom-up*, starting with the detailed design of the components and putting these together in an essentially additive way. Alternatively there may be a process which works predominantly from the *top-down*, in which decisions about the overall scheme are made first before the details of the parts are considered.

To illustrate this let's look at the Apollo programme of the 1960s. In 1961, President John F. Kennedy said:

> I believe that this nation should commit itself, before this decade is out, to landing a man upon the Moon and returning him safely to the Earth. We choose to go to the Moon! We choose to go to the Moon in this decade and do the other things, not because they are easy but because they are hard, because that goal will serve to organize and measure the best of our energies and skills, because that challenge is one that we are willing to accept, one that we are unwilling to postpone, and one which we intend to win...This is in some measure an act of faith and vision, for we do not know what benefits await us...But space is there and we are going to climb it.
>
> John F. Kennedy (1961): *Special Message to the Congress on Urgent National Needs*, Washington, D.C., 25 May

Exercise 2.2

Consider Kennedy's speech as an example of a design brief, such as outlined in Step A in Table 2.3.

(a) What broad objectives did Kennedy set?

(b) What constraints did he identify?

(c) What criteria did he set for success?

So what did NASA do the next day? Of course, NASA was already prepared for Kennedy's challenge: a politician would not make such a statement if the engineers were likely to say it was impossible. NASA mission designers were already able to estimate quantitative mission needs and requirements – Step B in Table 2.3. NASA had begun its Mercury programme of launches, and was developing plans for later Mercury and Gemini missions. These would include such operations as docking (linking) two spacecraft in orbit, investigating the effects of long-duration spaceflight on astronauts, and developing the technology of communicating with spacecraft in 'deep space'. NASA was also able to draw upon studies of alternative mission concepts – Step C of Table 2.3 – some of which were first formulated by engineers almost 20 years beforehand.

Before Kennedy's speech the decision had been made that the Apollo project would be based on 'Lunar Orbit Rendezvous' (LOR), with the mission consisting of two spacecraft, one of which would descend to the Moon's surface and then return to rejoin the 'main' craft. These two craft would separate and rejoin while orbiting the Moon. Design options that were discarded were:

- direct ascent of a single craft from the Earth to the Moon and back;
- Earth-Orbit rendezvous, with the separation and docking being performed in orbit around the Earth.

This is an example of a top-down decision, early in the programme, based on a great deal of design and analysis work done previously. When this decision was made, there was no plan for the design of the rocket motors, or the control panel of the Moon lander.

Why was this option selected? In terms of energy expenditure, LOR was calculated to be the most economical option. It called for the following steps.

1. Ascent from the Earth to a 'parking orbit' around the Earth.
2. Transfer from Earth Orbit to Lunar Orbit.
3. Descent by two astronauts to the Moon's surface in a Lunar Excursion Module (LEM) while a third astronaut remained in Lunar Orbit aboard a Command and Service Module (CSM).
4. At the conclusion of surface activity on the Moon, the explorers would return to lunar orbit in the LEM's ascent stage – the LEM itself would split into two sections, with the heavy landing motors remaining behind on the Moon's surface.
5. The LEMs ascent stage would then rendezvous and dock with the Command and Service Module for transfer of the astronauts from the LEM ascent stage. The return leg would be made in the Command and Service Module, and the LEM's ascent stage would be undocked and jettisoned. The astronauts' capsule would land in the ocean.

The 'top-down' LOR decision led in turn to a decision to build two separate manned modules for different phases of the mission, with these modules flying together as an 'assembly' during the Earth–Moon and Moon–Earth legs of the mission.

NASA mission designers also had calculations that showed that the mass that would need to be delivered to the Moon – about 50 tonnes – was within the lifting capability of the Saturn V launcher already under development by NASA's German-born rocket engineer, Wernher von Braun, and his team in Huntsville, Alabama.

The first of the 'Saturn' family of launchers had already been used successfully in 1961, and confidence was growing that incremental development of the Saturn family would lead to a rocket powerful enough to launch Apollo missions to the Moon by the end of the decade. Here again a 'top-down' process was evident. Learning from the experience of the Saturn 1 launch, von Braun and his team set a realistic upper limit on the masses that could be launched from the Earth's surface and boosted towards the Moon. The Apollo spacecraft (CSM and LEM) designers had to design within these constraints.

In contrast with top-down design, bottom-up design tends to be found at the detailed design level of individual components. If a designer is working on a new type of kettle, but has been told to use standard parts to reduce costs, then the design process is constrained by the pieces that have to go into the product, rather than by a top-down decision about performance or styling. An industrial designer might sketch out the concept for a new kettle shape, and this new shape might be seen as top-down design, because the product must look like the designer's concept. The bottom-up design starts when a heating element, power cable, safety devices and polymer mouldings have to be designed in order to make a functioning kettle and to satisfy the shape generated by the initial top-down design stage.

So in comparing the top-down and bottom-up approaches, it is important to realize that neither approach can succeed without the other. They never work in isolation, and both will be found in most projects. Thus a high-level decision made about an overall design scheme depends crucially on the form, properties, function, and availability of lower-level components. Indeed, a high-level design specification may be impossible to realize for various reasons, for example because it violates the laws of nature, because it is too costly or too complicated, or because the 'necessary' components have not yet been invented or developed.

Similarly, low-level decisions made about the detailed design of components or their interconnection in systems depend crucially on the requirement for reliability of function, and constructability of higher-level subassemblies and assemblies. Low-level design will not by itself prevent interference or adverse interaction between components; nor will it indicate mutual incompatibility between subsystems. Also, low-level design by itself will not determine the suitability and reliability of the system as a whole.

One further characteristic of these alternating phases of synthesis and analysis in design is that they tend to move from the general and tentative (top-down) to the more specific and definite (bottom-up). At first the designer considers preliminary outline ideas or sketch proposals. The evaluations made here are perhaps of a rather informal and approximate kind; later the preferred designs are fleshed out in more detail and analysis of performance and cost are made with appropriate precision.

So it's almost always some combination of top-down and bottom-up.

4.2 Managing design

Now that we've established that space projects are inherently complex, I'm going to look at three major organizational forms that can be used to house engineering projects. I shall examine the advantages and disadvantages of each of them, and discuss some of the critical factors that might lead us to choose one form over the others.

4.2.1 The functional organization

Figure 2.14 illustrates a purely functional organization. Projects will be allocated to a department on the basis of the perceived function of the project. So if the project in question involves a new technology, it will be placed under the supervision of the Director of Engineering. If it involves the introduction of a new product we would be more likely to find it under the control of the Director of Marketing. If, on the other hand, the project concerns the installation of a new manufacturing process it would probably be overseen by the Director of Manufacturing. A new project tends to be assigned to the functional unit that has most interest in ensuring its success or that can be most helpful in implementing it.

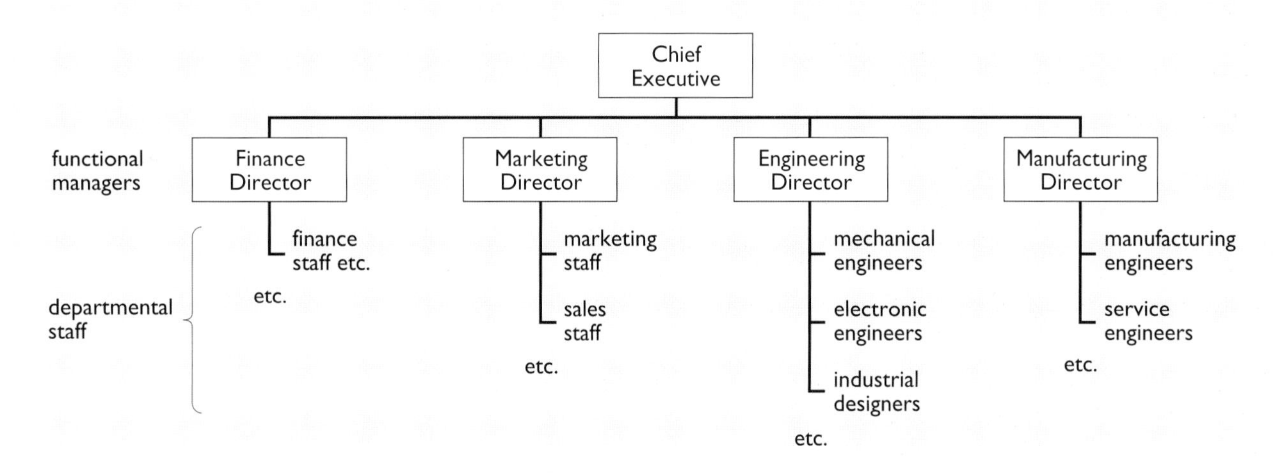

Figure 2.14 A functional organization

There are advantages and disadvantages of using functional divisions of the parent organization as the administrative home for a project, assuming that one has been able to find an appropriate administrative home for the project.

One advantage is that an engineering design or a manufacturing project will be free from 'interference' by sales staff. But this could be a disadvantage also. The engineering team may be working on a project which the marketing department would consider to be impossible to sell. There may also be insufficient input from the financial side of the organization.

Another advantage is that there should be good cross-fertilization of ideas between projects within departments, as each department will probably be working on several projects at once.

4.2.2 The project organization

At the other end of the organizational spectrum is the pure project organization (Figure 2.15). The project is separated from the rest of the parent organization. It becomes a self-contained unit with its own technical staff, its own administration, and tied to the parent firm perhaps only by periodic progress reports. Some parent organizations prescribe administrative, financial, personnel and control features in detail. Others allow the project almost total freedom within the limits of final accountability. There are examples of almost every possible intermediate position.

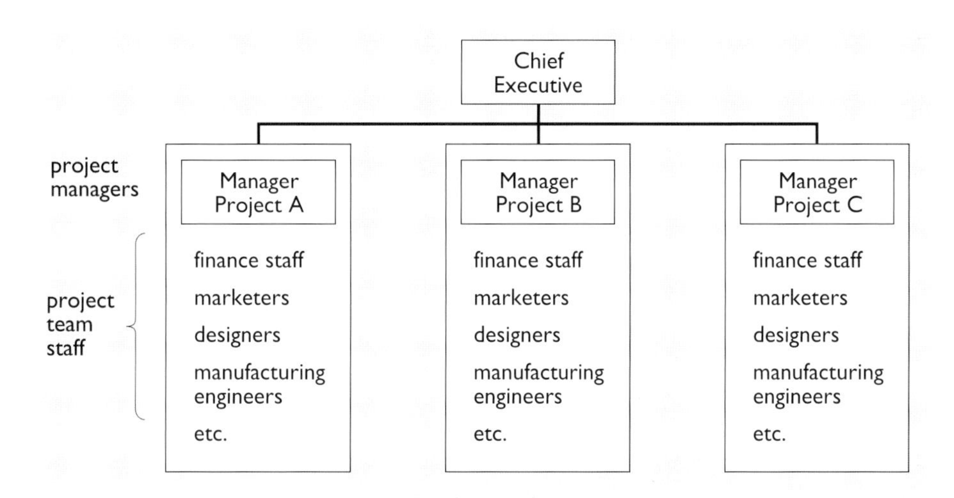

Figure 2.15 A project organization

As with the functional organization, the pure project organization has its advantages and disadvantages.

An advantage is that there is no competition between projects for staff time: staff will be assigned to one project only. This means, though, that there will be little scope for transfer of innovation between projects: staff working within one company will not necessarily be talking to each other.

Another disadvantage is that of accountability to the parent firm. A weak project manager may cause failure of a project, if the failings are not realized in time.

4.2.3 The matrix organization

In an attempt to couple some of the advantages of the pure project organization with some of the desirable features of the functional organization, and to avoid some of the disadvantages of each, the matrix organization was developed in the 1960s by the aerospace industry and in particular by the space industry. In effect, the functional and the pure project organizations represent two extremes. The matrix organization is a combination of the two. It is a pure project organization overlaid on the functional divisions of the parent firm.

A matrix organization can take on a wide variety of specific forms depending on which of the two extremes (functional or pure project) it most resembles. Because it is simpler to explain, let us first consider a *strong matrix* organization, which is similar to a pure project structure. Rather than being a stand-alone organization, like the pure project, the *strong matrix* organization is not separated from the parent organization (Figure 2.16). Let's investigate how this works.

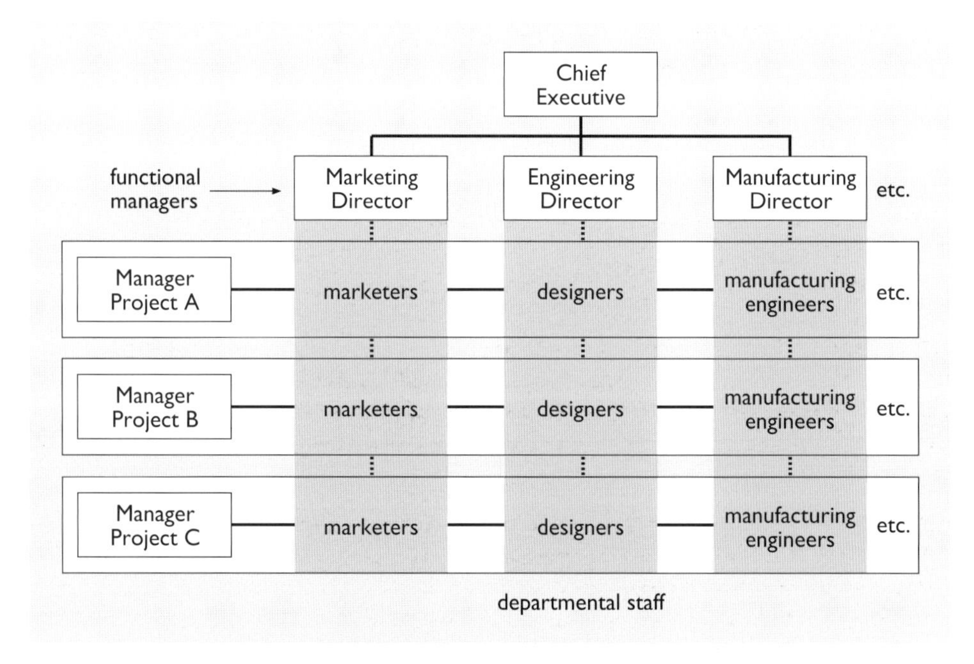

Figure 2.16 A 'strong matrix' organization

Project A might have assigned to it three people from the manufacturing division, one-and-a-half people from marketing (that is, one person full-time, and one person half-time), one-half of a person each from finance and personnel, four people from R&D (research and development), and perhaps others not shown. These individuals come from their respective functional divisions and are assigned to the project full-time or part-time depending on the project's needs. The project manger controls when and what these people will do, while the functional managers control who will be assigned to the project and what technology will be used.

The project managers may report to the head of the functional department with the major interest in the project, or the project manager of each project might report to a 'chief' project manager who also exercises supervision over other projects. If the organization conducts relatively few projects, there may be no need for a 'chief' project manager. Instead the project managers may report to the head of the functional department with the major interest in the project.

At the other end of the spectrum of matrix organization is the *weak matrix*, which is more like the functional organization (see Figure 2.17).

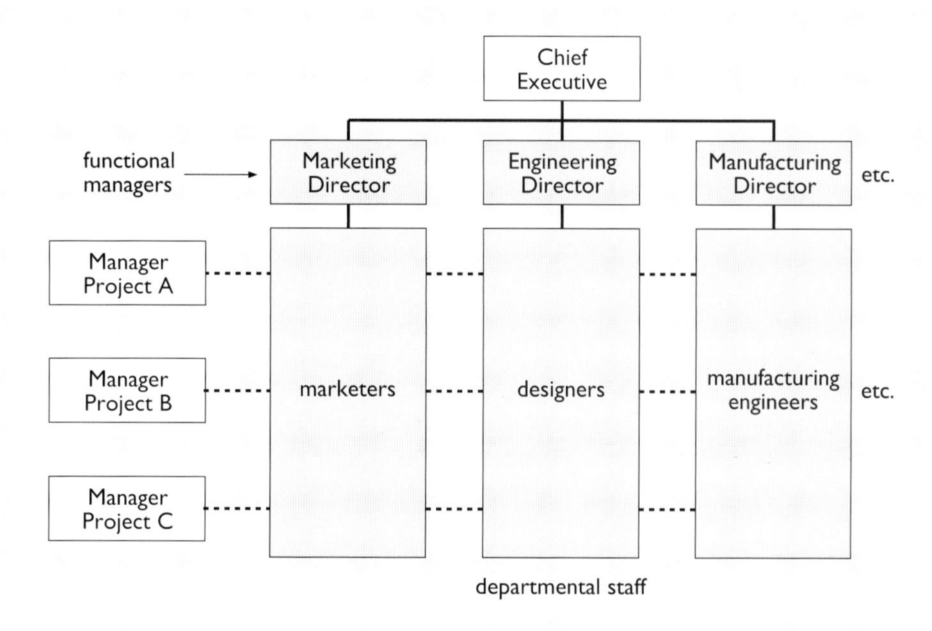

Figure 2.17 A 'weak matrix' organization

A project might now have only one full-time person: the project manager. Rather than having functional specialists assigned to the project, the functional departments may be asked to take on specific jobs required by the project. For example, the manager of a project to create a new nozzle for a rocket engine might request that the basic design be done within the engineering group, and the outcome passed to the manufacturing group. The priority given to the design might be assigned by senior management, or might be the result of negotiations between the project manager and the head of the design or manufacturing group.

The impetus for the matrix organization was the fact that firms operating in high-technology areas had to integrate several functional specialities to work on a set of projects, and there was a wish to share expertise between individual projects in the set. Further, the technical needs of the projects often required a systems approach; looking at each component separately was a drawback. In earlier times, when a high-technology project was undertaken by a firm it would start its journey through the firm in the R&D department. Concepts and ideas would be worked out and the result passed on to the engineering department, which would sometimes rework the whole thing. This result would then be forwarded to manufacturing, where it might be reworked once more in order to ensure that the output was manufacturable by the firm's existing machinery. All this required a great deal of time and the developing project might bear scant resemblance to the original specifications.

In the meantime another firm would be doing much the same thing on another project for the same customer. These two projects might later have to be joined together, or to a third, and the combination was then expected to meet its intended function. For example, the first might have the Apollo Command and Service Module, the second the Lunar Excursion Module (LEM), and the third the Saturn IVB upper stage. It would have been very unlikely that the composite result would have worked if all three developments went on in isolation.

The systems approach was adopted as an alternative to the traditional method described above. So the management of the overall project had to have a clear idea of what was the system that they were developing, its component subsystems and, just as importantly, the interrelations between these subsystems. This did not mean that the same firm had to manufacture everything, but it did mean that one organization had to take responsibility for the integrity of the overall design – to make sure that the parts were compatible and that the combination would function as expected. It was important that the designers and manufacturers of the component subsystems also had a view of the whole system which included the relevant interconnections to other subsystems, and a conception of the overall project.

4.2.4 Choosing an organizational structure

Even experienced practitioners find it difficult to explain how one should proceed when choosing the organizational structure for a project. The choice is determined by the situation, but even so the decision is partly intuitive. There are few accepted principles of design, and no step-by-step procedures exist that give detailed instructions for determining what kind of structure is needed and how it can be built. All that can be done is to consider the nature of the potential project, the characteristics of the various organizational options, the advantages and disadvantages of each, and then to make the best compromise.

In general, the functional form is the best choice for projects where the major focus must be on the in-depth application of a technology rather than, for example, on minimizing cost, meeting a specific schedule or achieving speedy response to change. Also, the functional form is preferred for projects that will require large capital investment in equipment or buildings of a type normally used by the function.

If the firm engages in a large number of projects in one field (e.g. construction projects), then the pure project form of organization is preferred. The same form would generally be used for one-time, highly specific, unique tasks that require careful control and are not appropriate for a single functional area – the development of a new product line for instance.

When the project requires the integration of inputs from several functional areas, and involves reasonably sophisticated technology, but does not require all the technical specialists to work for the project full-time, the matrix organization is the most satisfactory solution. This is particularly true when several such projects must share technical experts. Matrix organizations are complex, though, and present a difficult challenge for the project managers. They should be avoided when simpler organizational structures are feasible.

SAQ 2.7 (Learning outcome 2.3)

(a) A project to develop a new rocket-fuel storage system requires input from an engineering design team, with advice from chemical engineers and manufacturing engineers. Design and manufacturing engineers are needed to work on the project full-time, whereas the input from the chemical engineer can be part-time. Some of the engineers on the project have expertise which is appropriate for the development of an associated project for a different launcher design. What is the most appropriate organizational type for this project?

(b) Having developed the system, the controlling firm wishes to outsource its manufacture. The company that will build the device will receive a detailed, fixed design specification, including all materials and dimensions. What is the most appropriate organizational type for the firm that will build the device?

4.3 The vicious and virtuous circles of designing for space

For many years space projects have been promoted and carried out under the assumption that 'space is expensive'. But once we assume that space is expensive we enter into 'vicious circles' of decision making that *guarantee* that it will be expensive. In 1992 Dan Goldin, the NASA Administrator, challenged international space agencies, industry, and universities to break away from the old ways of doing things and enter 'virtuous' circles of 'Faster, Better, Cheaper'.

4.3.1 Expensive and too-expensive space projects

We have seen how President Kennedy's 1961 speech outlined the goal of travelling to the Moon. For the Apollo programme of the 1960s, considerations of cost were secondary to the focus of achieving the national goal set by the President.

Twenty years after the first Moon landing President George Bush tried to emulate President Kennedy:

> The time has come to look beyond brief encounters. We must commit ourselves ... to a sustained programme of manned exploration of the Solar System and the permanent settlement of space.
>
> George H. W. Bush (1989) *Remarks on the 20th Anniversary of the Apollo 11 Moon Landing*, Washington, D.C., July 20

Several plans for a manned mission to Mars were drawn up in the US. One such plan had a cost estimate of 450 billion US dollars (1989 prices). Bush's programme quickly became doomed. Taxpayers did not want 'money-no-object' programmes, not even programmes that would drive technology forward and would have political benefits. Instead, 'value-for-money' projects and programmes in the space domain were the order of the day: space projects had moved ▼From technology push to market pull▲ (p. 128). Taxpayers, particularly US taxpayers, are interested in low taxes; meaning that projects need to be cost-effective, and preferably have a commercial potential.

4.3.2 Changing the paradigm

Evidence that cost reduction was possible came from several directions. For 30 years the then Soviet Union launched many more satellites annually than the US, but with roughly the same budget. The capability of these satellites was limited, and consequently the 'cost per unit of performance' may have been comparable to those launched in the West. However, the number of spacecraft launched and their total mass far exceeded the totals launched by the West. These facts implied a cost per kilogram far lower than the United States and Western Europe had achieved.

For many years university and other research groups had developed ever more capable satellites at a cost of the order of one-tenth that of traditional

▼From technology push to market pull▲

'Technology push' and 'market pull' were two early models of the innovation process.

Technology push was dominant in the 1950s and 1960s. It is a simple linear model which suggests that the innovation process starts with an idea. Sometimes this idea comes to a creative individual who has the knowledge and imagination to realize its significance, and the practical skills to transform the idea into an invention (Figure 2.18). However, more often nowadays the starting point is basic scientific research or applied R&D in organizations. This then proceeds through design and development into a product which can be manufactured effectively and economically and then sold on the market.

The market is seen as a receptacle for the output of scientific research and invention.

Although the technology push model might describe the innovation process for some products, it only tells part of the story. There are numerous examples of inventions which are good ideas scientifically, or technologically sound, and available to the market, yet which fail to become successful innovations. The notion that if an idea is good enough the 'technology push' will help it overcome all obstacles to its innovation is a romantic one, but unrealistic!

As a reaction against the 'technology push' model the 'market pull' model came into prominence in the 1960s and 1970s (Figure 2.19).

The market pull model suggests that the stimulus for innovation comes from the needs of society or from a particular section of the market. According to this model, a successful approach to innovation would be to research the market thoroughly first: assess what needs exist; assess how far they are met by existing products and processes, and how the needs might be met more effectively by means of a new or improved innovation. The theory is that once the appropriate technology is developed, a receptive market is assured because the innovation process has been tailored to meet a definite need. Thus this model adds a stage of exploring market need *before* the invention stage of the technology push model. This approach has been characterized by the classic saying 'Necessity is the mother of invention', or as expressed by one writer in this field 'Find a need then fill it'.

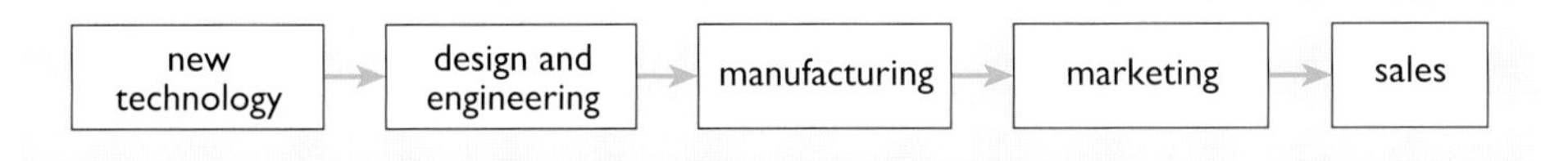

Figure 2.18 The model of 'technology push'

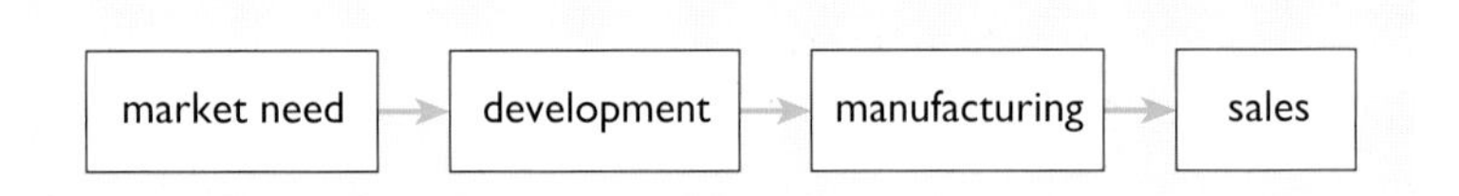

Figure 2.19 The model of 'market pull'

programmes. Also, Soviet and Western launchers developed for small commercial satellites had equal or better reliability than those from traditional, large programmes.

In the 1970s and 1980s a wide variety of *incremental* improvements in cost reduction were applied to large government and commercial satellite procurements. These methods such as **▼Design to cost▲** and **▼Concurrent engineering▲** were seen to have the potential to drive down costs by 10% to 50% as they had done in other industrial sectors.

Radical cost-reduction methods offering the possibility of reducing costs by a factor of 2 to 10 were also tried by the innovative companies in the space sector. These methods sought to change the paradigm by which satellite acquisition worked; they were inconsistent with traditional methods and sometimes with each other. In almost all cases radical cost reduction required changing the rules of the game. No longer would 'user requirements' be the sole driver of the design process. Rather there would be 'trading on requirements' – that is, the overall mission objectives would be approached by developing compromises between what the users said they wanted, and what was available at low cost.

Design to cost

'Design to cost' tries to estimate the cost of components and subassemblies. The method then compares costs of alternatives to determine which is the lowest cost design. This method does not necessarily require the 'actual' cost of items, but certainly needs the relative cost of one alternative as compared to another.

A design-to-cost analysis can be undertaken either narrowly or broadly. A narrow analysis looks at how much it costs to make a product, and is intimately related to the price paid by a customer. A broad analysis also takes into account the costs of owning and operating (and maybe also disposing of) the product after it is purchased. These are the 'whole life' costs of the product.

Narrow analysis

In this analysis,

Cost = design cost + raw materials cost + cost of manufacturing

Manufacturing costs include time on machines and labour to manufacture components and subassemblies. The costs may be standardized to a cost per hour, irrespective of the type of manufacturing task. However, this is very inaccurate. It is better to match a current component or subassembly as closely as possible to a similar design which has already been produced. If the cost of the existing item is known, it is possible to estimate the cost of the new item. This depends on historical data on costs being available. The main elements in applying 'design to cost' are (i) creating a database of costs and (ii) matching new designs to similar items in the database.

Broad analysis

In this analysis,

Cost = design + manufacture + raw materials + operating + maintenance + disposal

The idea here is to include costs to the customer other than just purchase costs. Perhaps more significantly, the cost is also intended to include the wider environmental costs incurred by a product. For example, pollution has a cost both to the user and to other people. These costs are often hidden but are nonetheless real for the general population in terms of taxes and effect on the quality of life. So a new design of car engine might cost more to produce, but could be cheaper to run and produce less pollution over its lifetime. An analysis like this is often called 'life-cycle analysis'.

Neither traditional (incremental) nor radical cost reduction was seen as inherently better in all situations. Which approach to apply, or even whether any major attempts at cost reduction were appropriate, was seen as dependent upon the nature of the project in question. After all, the processes of space mission design and development were not painstakingly built up by the space community over 30 years in isolation or without any concern about cost. Even governmental coffers are not bottomless! Almost any attempt to reduce costs introduces some elements of compromise.

4.3.3 The vicious circle of designing for space

Why do space projects cost so much? Perhaps the problem starts with the commonly held belief that space projects are expensive, which limits their number. Only having a few projects means:

- they must be planned carefully to get the most out of them;
- they must be reliable;
- they need to be large to achieve a lot from each opportunity;
- there will be little competition (of ideas or between suppliers) if new projects start infrequently.

Each of these statements brings consequences for conducting the projects.

Extensive planning often leads to delays. A requirement for high reliability means that integrity must be achieved by design. This precludes using the latest unproven technology and this, taken with the planning delays, means that spacecraft are built with obsolete parts. Large payloads need large spacecraft which in turn need large launch vehicles. When combined with the lack of competition, and the need for high reliability, this means that launches become expensive.

Concurrent engineering

This method of reducing project costs is sometimes called *simultaneous engineering*. It is an example of a multidisciplinary approach to complex projects.

Conventional ways of organizing projects had tended to divide a large project into separate parts with very little communication (think about a functional organization from earlier, where a project in a particular area has further subdivisions within the group). Specialists dealt with each part. The danger in this approach is that although individual parts will function excellently in their own right, they may not work together effectively. There are two consequences. Unnecessary costs will have been incurred both in designing and in putting the final product right. Further, the overall time to deliver the product to customer is increased and the quality (product performance) is likely to be poor. A picture of this situation might be as shown in Figure 2.20.

The whole task is divided into separate tasks. After the specifying stage, the product is divided into subsystems. Three of these, Navigation, Power and Structure, are indicated in the diagram. The arrows represent the order of the tasks. Specifications, drawings and subassemblies are passed between stages. Each task is completed before moving the next is begun. The drawbacks here are that (i) unresolved problems and mistakes are passed on to the next task and (ii) new problems arise at later stages because of 'arbitrary' decisions made at earlier stages. Both lead to extended delivery times and poor quality.

Concurrent engineering takes a team approach, trying to increase communication among tasks. The product is considered as a whole product rather than a collection of separate parts. A picture of what happens when concurrent engineering is working well might be as shown in Figure 2.21.

Notice the overlap between tasks. The arrows indicate channels of communication. This overlap reduces time to deliver the product, improves communication and increases the likelihood that all parts of the product will work effectively together.

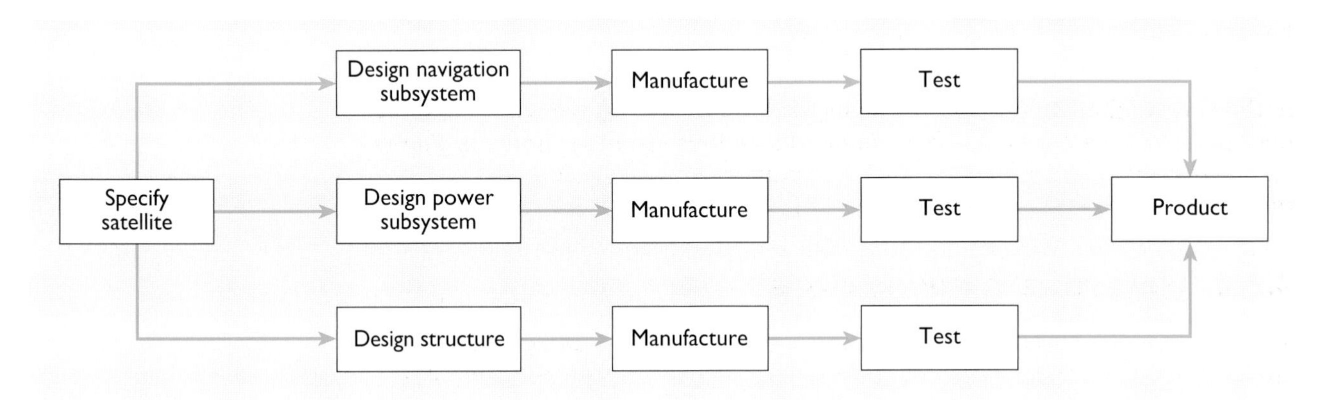

Figure 2.20 Conventional engineering approach. For each subsystem, design, manufacturing and testing happen sequentially

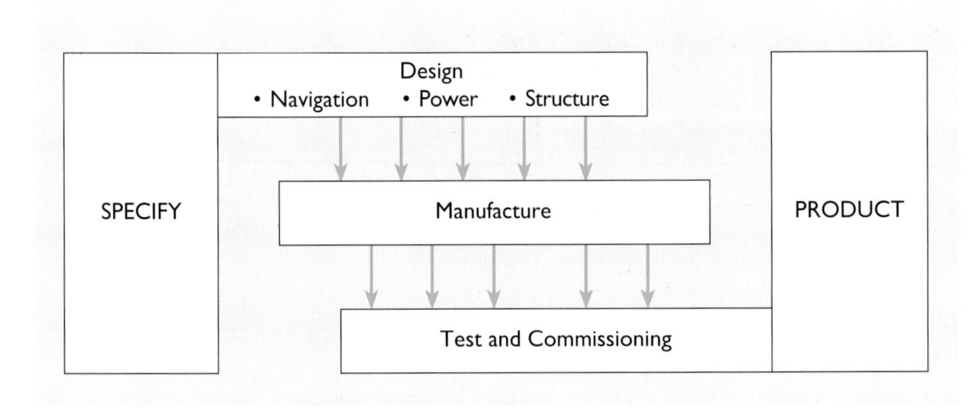

Figure 2.21 Concurrent engineering

Large spacecraft need large (and expensive) ground infrastructure. As the space and ground systems become too large to be built by a single contractor, so management structures become complex and expensive. As a result, trade-offs to optimize parts of the system become difficult because of the rigid contractual and organizational boundaries.

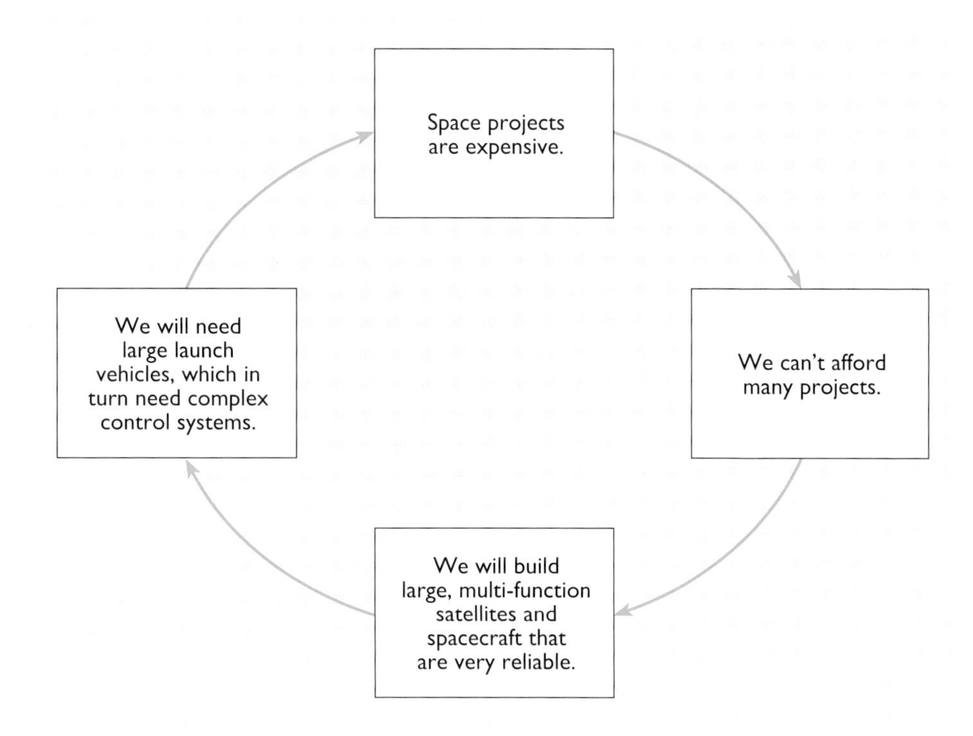

Figure 2.22 The vicious circle of space design

In the end we don't get the best designs, and often have to live with relatively poor performance per kg. We have entered a vicious circle as shown in Figure 2.22, whereby poor performance per kg requires even bigger payloads reinforcing the opening premise – 'space projects are expensive'.

4.3.4 Towards a virtuous circle

Several space business failures illustrate very clearly that space is a difficult area for private enterprise. A recent example at the time of writing (2000) is the Iridium satellite telephone system, which used a network of 66 satellites to provide global coverage for a phone linked to its network. The business failed, though, to achieve its targets for income, which was attributed to poor marketing and pricing rather than problems with the technology.

It's equally difficult for the public sector to find economic justification for space exploitation, with the relentless pressure to reduce spending and seek more cost-effective ways of achieving national goals. How might these economic pressures affect engineering decisions? We can expect five key changes:

- space agencies will be forced to establish priorities based on user needs (market pull again);
- failures will be more tolerated in smaller, less expensive missions;
- appropriate technology will be used;
- missions will be dedicated to fewer payloads;
- projects will have a shorter duration and be developed by a smaller integrated team.

Let's examine each of these to see the implications.

- User needs. The essential difference between a project driven by a space agency and one driven by a user is the measure of success. Space agencies tend to measure success by inputs such as the size of the budget, advancing technology, or completing the project, not by how well it satisfies users' needs. Users measure success by outputs such as the quality and quantity of data or the service they receive.

- Failures tolerated. Being competitive often involves taking calculated risks and accepting some failures. Insisting on perfection in space systems is fundamentally wrong. The perfect system, whether it is a spacecraft or a toaster, will never be built.

  The description of the vicious circle above concluded that everyone behaves reasonably when making the individual decisions that collectively lead to such an unreasonable consequence. To achieve the acceptance of failure we must change the motivation that drives each decision. As David McLelland, a business researcher, said in the 1950s:

  > If the penalty for failure exceeds the reward for success then people will plan not to fail, not plan to succeed.

- Appropriate technology. It's normal to demand full space qualification for all components and subsystems, but this decision should depend on how critical they are to the mission and the stress they must undergo. By choosing specifications properly, designers can use reliable, well-engineered products, while avoiding the expense of procuring special purpose items.
- Missions dedicated to fewer payloads. Large multi-payload satellites are like a convoy only able to move at the speed of the slowest ship. If any part is delayed the overall project is delayed.
- Projects with short duration and a small team. Space technology often lags by ten or more years because a long duration project requires an early specification of its design, which then cannot be evolved further as the product is also the prototype. For a complex project with a large team, an early specification is necessary to define the interfaces between elements of the spacecraft, so that the distributed team can start their detailed design work.

  Project costs go up proportionally with duration and team size. Teams are usually kept working and spending during project delays, and complex management interfaces means people spend a lot of their time generating documentation. Wernher von Braun spotted this in the 1950s:

  > We can lick gravity, but sometimes the paperwork is overwhelming.
  >
  > Wernher von Braun (1958) *Chicago Sun Times*, 10 July.

Project management and changes driven by management represent the largest cost element for any projects, especially if international collaboration is involved. Small teams, preferably at one location, working intensively on a short project can bring this cost down.

What would happen if all these changes in approach were encouraged? If we assume that space projects are cheap, a different positive feedback loop emerges: a virtuous circle as shown in Figure 2.23.

This is the rationale behind the 'Faster, Better, Cheaper' missions developed by NASA at the end of the 20th century. The loss of two Mars probes, during the 'Faster, Better, Cheaper' regime, although embarrassing and costly, has not destroyed NASA's entire space exploration programme.

SAQ 2.8 (Learning outcomes 2.9 and 2.10)

Two possible exploration programmes to Mars are:

1. A mission where a team of astronauts visit the planet for exploration and research;
2. A series of missions using unmanned orbiting satellites and landing probes.
   - (a) Classify each of these in terms of the vicious and virtuous circles of space design. Justify your answer using the boxes in Figures 2.22 and 2.23.
   - (b) What changes in the funding and design of space missions have made option 2 more favourable than option 1?

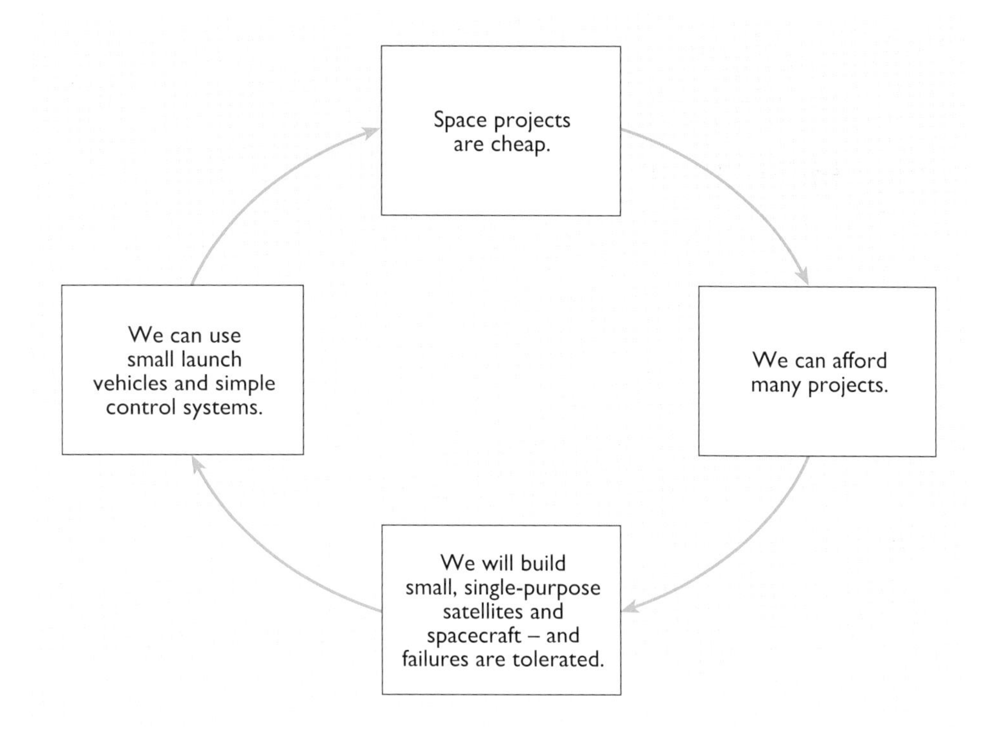

Figure 2.23 The 'virtuous circle' of designing for space applications

What this discussion illustrates is that the approach to a design problem can be affected by constraints such as budget and prejudicial expectation of what the final project will be. Reductions in budget, or in broader terms, in market price for a product, will influence the design of the product inasmuch as it will have to become cheaper and yet still be viable. Changing the paradigm of space missions to assume that they *will* be cheap and some *might* fail, has allowed for large changes in the design of such missions and their overall cost.

We will look further at the influence of economics in the next section.

5 The business of space

5.1 Fundamental principles

Classical economics considers the supply and demand for products in the marketplace. It assumes:

- there are many customers and many suppliers,
- products (goods or services) are well-defined and well-understood,
- there is rational behaviour on all sides.

With these assumptions, we can understand some basic economic concepts. The most important of these is the relationship between supply and demand and its effect on price.

The classical theory of supply and demand says that as the price of a product rises, more and more sellers will want to enter the market (to make money from sales of the product), but customers will want to buy fewer units of that product (because it costs too much). An efficient market will establish an equilibrium price at which the quantity supplied equals the quantity bought. I show this graphically in Figure 2.24. The equilibrium price is where the supply and demand curves intersect. You may have been aware of this approach in the electricity business. There are schemes that sell electricity more cheaply during off-peak periods: you may have, or know someone who has 'Economy 7' storage heaters, which are heated using electricity in off-peak periods, and which then release heat slowly throughout the day. Such schemes recognize that customers need to be given an incentive to buy electricity during off-peak rather than peak hours, so the electricity companies stimulate demand in off-peak hours through price reductions: they can then keep many generators running efficiently near full capacity all day.

Changes in circumstances will affect the behaviour of the suppliers and customers. Imagine, for example, the impact of a heat wave on the market for

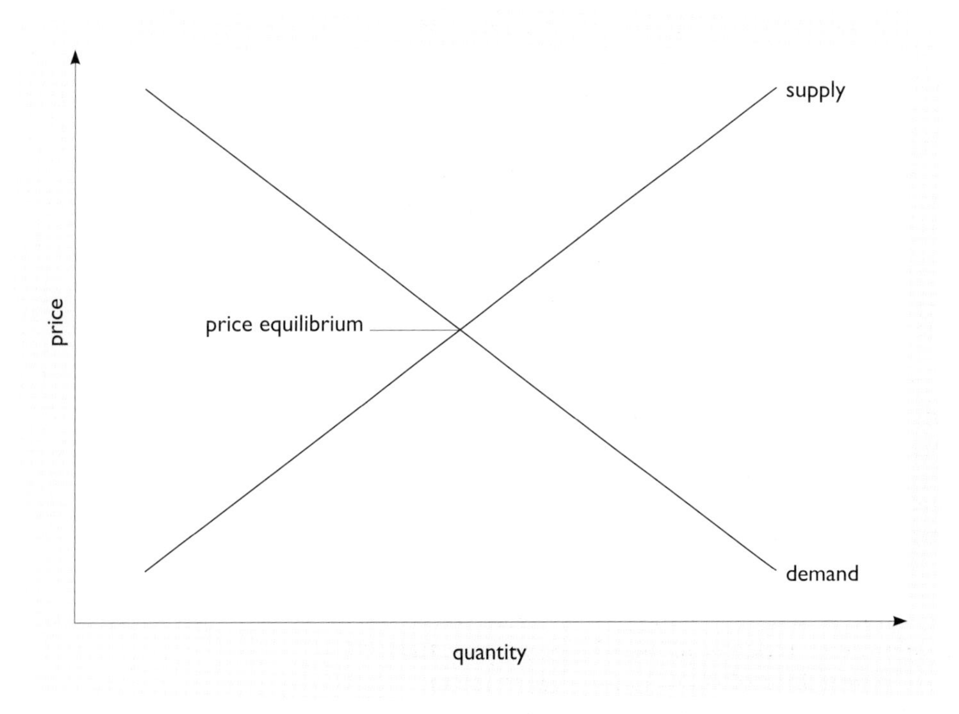

Figure 2.24 Price, supply and demand

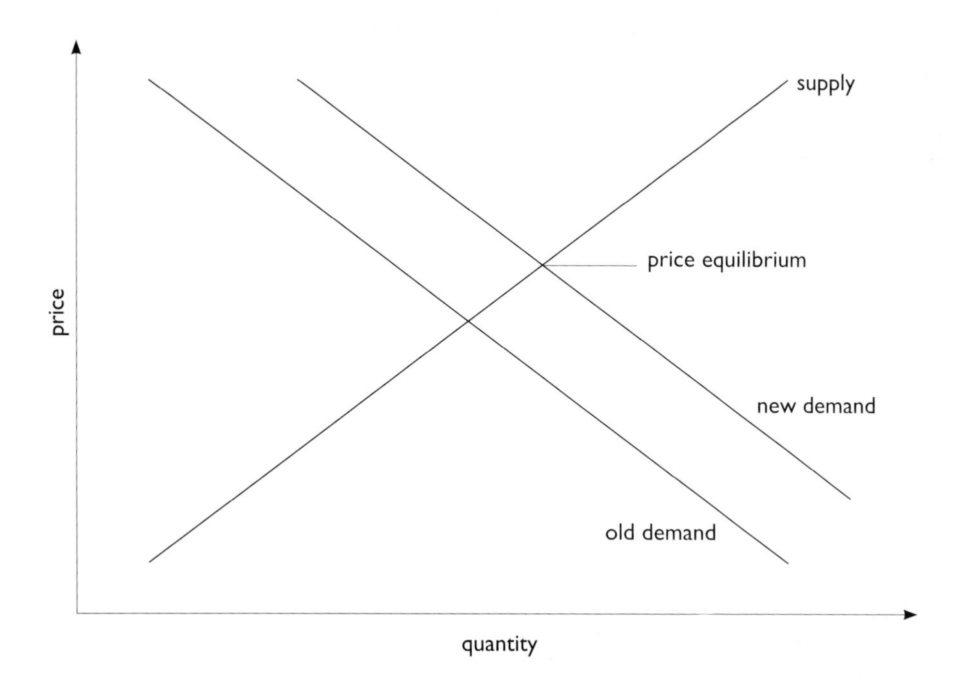

Figure 2.25 The effect of increased demand

cold drinks. Suddenly, more customers want to buy, so the demand curve moves outwards, and the equilibrium price increases as shown in Figure 2.25.

If the bottling plants are now running at peak capacity, suppliers may bring into service additional production lines to meet the higher demand. These lines are probably less efficient (which is why they were closed in the first place), and so the stabilization of the price at a higher level than before the heat wave is not necessarily simply a result of the suppliers taking advantage of higher sales to increase their profit margin.

Of course, in the real world the effects of delays must be taken into account. By the time additional production lines become operational, the weather may have changed and the demand returned to normal. The system could become unstable if there are large and rapid fluctuations in the supply, demand and price of the product. No wonder electricity prices are set over a longer term: can you imagine trying to regulate your usage in response to a daily price adjustment?

Does the simple supply-and-demand-price model work in space? Look back at the first two assumptions of the model: many customers, many suppliers, and well-defined and understood products. These don't yet apply to the space domain. The classical theory applies to what economists call 'commodities' – goods and services that are essentially the same wherever they come from and where customer choice is often based on price.

To better understand the space domain, I will introduce a more complicated economic model which applies to the development of high-technology products. Unlike commodities in the classical supply and demand model, the products I have in mind get cheaper as the number of units sold increases. Take for instance, pocket calculators. When it was first introduced, a pocket calculator cost a week's wages. Now a basic model can be obtained for less than an hour's pay.

This effect arises as a consequence of the *learning curve* and *economies of scale*. The *learning curve* means that the expertise obtained in developing the first few prototype units allows the production of subsequent ones at lower cost. *Economies of scale* allow cheaper mass production and sharing of development costs over larger production runs to decrease the cost per unit.

▼Globalstar▲

The Globalstar consortium of leading international telecommunications companies was originally established in 1991 to deliver satellite telephony services through a network of exclusive service providers. The plan was to provide high-quality satellite-based telephony services to users outside the normal terrestrial coverage of land-based cellular systems.

The scheme is based on a constellation of 48 Low Earth Orbiting (LEO) satellites. As a wholesaler, Globalstar sells access to its system to regional and local telecommunication service providers around the world. These Globalstar partners, in turn, form alliances with additional providers. By extending its reach, Globalstar complements terrestrial networks, rather than competing with them, opening up markets and bringing telephony to entirely new areas of the world.

An order for the preparation and launch of 48 satellites looks almost like a production-line task!

▼Arianespace▲

Arianespace, a commercial company incorporated in 1980, has a share capital of 300 million euros (around 200 million pounds sterling) and 53 corporate shareholders. It is based in Evry in France.

Its main business is the marketing, sales and production of the Ariane 4 and Ariane 5 launch vehicles and associated launch services.

Ariane 4 offers a payload capacity of 4.9 tonnes. By October 2000 the Ariane 4 technology had achieved 58 successful flights in a row.

The Ariane 5 vehicle was developed for heavier payloads. It is intended to evolve into a family of launchers that responds to the space-transportation challenges of the twenty-first century. Starting in 1999 with a 6.2 tonne payload lift capability, the Ariane 5's performance is expected to grow to 12 tonnes in the early years of the twenty-first century.

Unfortunately batch and repeat orders have been unusual in the space domain, but there have been some notable exceptions: see **▼Globalstar▲** and **▼Arianespace▲**.

Usually, a space system has been a one-of-a-kind development for which the price paid by the customer has essentially been whatever it costs the supplier plus some profit. There is, however, a significant trend in the space business which will lead towards economies of scale and the benefits of a supply-and-demand-based marketplace.

5.2 Manufacturing set-top boxes for domestic TV receivers

The following example illustrates the effect of economies of scale on price and profit. We will then contrast this example with an example of space-based manufacturing.

The current example, that of set-top boxes, was prepared at the start of the digital television revolution when many consumers were looking for technology to upgrade their existing equipment. A set-top box is a device that decodes digital TV signals, and converts them for display on a conventional 'analogue' set.

It is the year 2000. Imagine you start a company that makes set-top boxes (STBs) for the reception of digital television, and you decide to develop a new model. The market price appears to be around £200 for your kind of product. You start with two people working out the specifications and, by the time you have a prototype of the new model six months later, the team has grown to 10 people. You establish a production department and the staff rises to 20, who work for another year to set up the production line. After 12 months you might have spent almost £0.5 million and have manufactured the first 55 units.

Exercise 2.3

You can now make and sell STBs, but how much has each of these first 55 off the production line effectively cost you, averaging the first year's expenditure over the first year's production?

This is possibly a bit too pessimistic a calculation. During the final quarter of the year, 50 of the 55 units were produced, and the expenditure during this period was about £200 000. So this batch of sets actually cost £4000 each, showing that the cost per box is beginning to reduce as production gears up.

A year later, after over 5000 STBs have left the factory, they only cost you just over £100 each to manufacture and you are able to establish an operating profit from which to pay off the overdraft.

After another year, when the market has grown and you have shipped 35 000, they cost you £80 each, but you may now be forced to sell them for less than £200 each if competitors are bringing down prices.

We can plot the effect of economies of scale in this example using a *cash flow* chart. Cash flow is the difference between the amount of cash received (income) and paid out (expenditure) by a company during an accounting period. I have set up a spreadsheet to do this using some notional costings for labour and materials. Figure 2.26 shows the quarterly accounts and Table 2.4 show the production and sales data that I used. You can check and extend my results in Activity 2.2 on p. 138.

Table 2.4 STB production and sales data

	Year 1				*Year 2*				*Year 3*			
Quarter	*1*	*2*	*3*	*4*	*5*	*6*	*7*	*8*	*9*	*10*	*11*	*12*
Staff cost	2	10	15	20	20	20	20	25	25	25	30	30
Units manufactured		1	4	50	200	500	1000	4000	4000	6000	10 000	10 000
Cumulative production		1	5	55	255	755	1755	5755	9755	15755	25755	35755
Quarterly sales					40	500	1000	4000	4000	6000	10 000	10 000
Cumulative sales					40	540	1540	5540	9540	15 540	25540	35 540
Unit cost per quarter/£k		100.05	37.55	4.05	1.05	0.45	0.25	0.1125	0.1125	0.0917	0.08	0.08

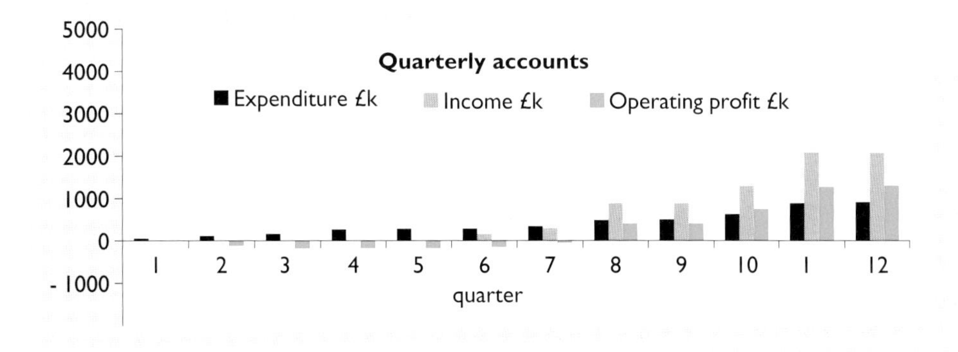

Figure 2.26 Quarterly accounts for the production of set-top boxes, using the data from Table 2.4 and the figures given in the text. Each quarter represents 3 months. A 'negative' profit means a loss!

Exercise 2.4

According to Figure 2.26, when does the company first make an operating profit?

Activity 2.2 (Learning outcomes 2.7 and 2.10) Making money from set-top boxes

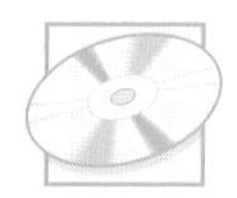

My spreadsheet for the set-top box example is one of the resources available on the T173 CD-ROM. Guidelines on using this spreadsheet are given in *Study Guide 2*, and you may wish to refer to this. You will not need to be able to write spreadsheets in order to perform this Activity: the spreadsheet is used as a tool to save working through detailed calculations.

The spreadsheet shows the quarterly accounts and balance sheet for the first few years. The costings that I used in the example were that a person costs £10 000 per quarter of a year to employ (£10 k/qr.), including the 'overhead' costs of providing them with an office and a computer etc.; and that each STB unit costs £50 for materials. There is an operating profit from the eighth quarter onwards (shown on the quarterly accounts), which leads to positive bank balances (shown on the balance sheet) from the tenth quarter onwards.

Exercises 2.5, 2.6 and 2.7 add a little more realism to this example.

Exercise 2.5 Payroll and price

(a) Change the figure for staff cost to £14k per quarter and note when the overdraft is cleared. (To change the figures, just click on the relevant box in the spreadsheet, type the new figure, and press the 'Return' key on your keyboard. You will then see the graphs change to reflect the new figures.)

(b) Now raise the unit price to bring the start of the positive balance back into tenth quarter. Don't overdo this! If you were to increase the unit cost too much, your sales would be hit, so find the unit cost that will maintain a small profit in the tenth quarter. Experiment until you find a suitable figure. The unit price is given in thousands of pounds (£k), so 0.2 corresponds to £200.

The spreadsheet allows the effect of competition to be modelled by reducing the unit price by a given percentage per quarter. You will almost certainly have noticed that new electrical and electronic goods tend to become cheaper with time: mobile telephones are a good example, and televisions with large screens become cheaper each year. Exercise 2.6 investigates this.

Exercise 2.6 Competition and price

With the staff cost reset to £10k/qr. and the base price reset to £0.20k/qr., include a 2.5%/qr. price reduction. How much does this delay the return to positive balance?

The spreadsheet also allows an interest charge to be levied by the bank on any negative balances. It is doubtful that a bank would not charge interest on an overdraft! However, a 'negative balance' on one product could be cross-subsidized from profits elsewhere in a company's range.

Exercise 2.7 Paying the cost of borrowing

Include interest charges at 20% and leave competitive price discounting at 2.5%. What new base price would enable the overdraft to be paid off by the tenth quarter?

▼Time is money▲

We need one more basic concept of economics to understand the space market, although it didn't have much influence on the space industry in the twentieth century.

The *time value of money* says that income in the future is worth less than income today (a monetary equivalent of 'a bird in the hand is worth two in the bush', if you like). Similarly, a payment that can be deferred is better than one that must be met today. In our daily lives, many of us apply this principle by using credit cards.

The quantification of this idea is achieved using an arithmetic procedure called *discounting*. Discounting is used to calculate the present value of future income and payments. An expected item of income, say five years in the future, is discounted (reduced) by a constant fraction for each intervening year back to the present to obtain a *present value* of that expected income.

A typical discount rate that a potential investor might use in a high-technology, high-risk project is 30–40% per year. The investor is essentially saying that an expected income in two years time is worth only around 50–60% of its apparent cash value today. Similarly, if the project can put off expenditure for a couple of years, it's only about half as painful as it would be today. So a project that is expected to be hugely profitable in 10 years' time probably would not be a great incentive for an investor to stump up the cash. But likewise, a piece of equipment requiring a major service in 5 years' time doesn't look too expensive on this year's balance sheet.

In Exercise 2.7, p. 138, the cost of borrowing is included in a business that establishes a steady income. We shall see what difference interest charges can make to projects where the income arrives at the end of an expensive space mission, along with the effects of several other risks, in the 'Made in space' case study in Section 5.3.

The cash flow of our STB manufacturer was calculated for a basic accounting period of a quarter of a year (see Figure 2.26 and Table 2.4). Notice that there is a negative cash flow (the expenditure is more than the income) for seven quarters, followed by an increasingly positive cash flow when serious production and sales start. A positive cash flow is important for the continued success of the company. In fact the cash flow must be sufficiently positive to pay off the bank loans quickly: ▼Time is money▲!

The problem in the space domain is that it has almost always been in the position of the first set-top boxes. As we saw earlier, the first five cost £30 000 each. It has been unusual to make two identical spacecraft; until recently the longest production line was ten or so that were similar (but not identical). Space is where the automotive industry stood before Henry Ford came along, said, 'You can have any colour as long as it's black,' and began to turn out identical vehicles by the thousand.

5.3 Made in space

In this section I'm going to bring together many of the issues of space engineering and economics to which you have been introduced in this part, and look at an imaginary commercial venture in space. By using realistic estimates of costs, risks and rewards, I will highlight how space ventures are different from other kinds of business, even though they are based on the same laws of economics.

5.3.1 The project

I am going to use in-orbit manufacturing as the example of a commercial space project. As the costs of launching materials into space and of retrieving manufactured products are high, the only hope of making a profit is when the final product has a high ratio of value to mass. Contact lenses are such a product. Depending on the type of lens, contact lenses can cost over £2 million per kg to the customer. This is because they weigh so little, and command a very high price to the customer in relation to their weight.

Is there a reason why space manufacturing might be good for contact lenses? Have a look at ▼Contact lens polymer in space▲.

▼Contact lens polymer in space▲

Contact lenses are available in several different types: soft lenses, gas permeable (or rigid) lenses and disposable lenses. Gas permeable lenses have several advantages over soft lenses: they are more durable (more resistant to wear and handling); they cause fewer problems for the health of the eye, as they are less prone to collecting bacteria and other contaminants; and they are generally considered easier to clean. They can also correct vision problems that cannot be corrected with soft lenses. Approximately one quarter of contact lens wearers need gas-permeable lenses. Gas-permeable lenses are made of 'gas-permeable polymeric material' (GPPM), which is a polymer that allows specific gases, such as oxygen, to pass through easily. A good supply of oxygen to the front of the eye is required to prevent swelling of the eyeball. The considerable drawback to these lenses is that they are not yet as comfortable for the wearer as soft lenses.

An experiment in manufacturing polymers for contact lenses in space was conducted in June 1993 aboard the Space Shuttle *Endeavour* in SPACEHAB. SPACEHAB is a pressurized module that is fitted in the Shuttle cargo bay and allows experiments to be conducted in a 'microgravity' environment: the environment that exists in an orbiting spacecraft where objects are essentially 'weightless'. The whole experiment had to be designed to fit in a locker providing a maximum capacity of about 27.3 kg (60 pounds) and about 0.06 m^3 (2.0 cubic feet) of volume.

The experimental module was a special polymerization module which contained a sealed aluminium container in which a series of test tubes were filled with 'monomers': the chemical precursors which are used to make the final polymer material. The polymerization module was then placed in a small combined refrigerator and incubator oven which could maintain temperatures over a range of 4 °C to 40 °C for indefinite periods. Twenty-eight different polymer monomers were placed into the flight module and maintained at 4 °C until the start of the experiment. A series of identical materials were placed in a second polymerization module and refrigerator/oven on Earth, to give a 'control' experiment to which the results of the space module could be compared.

One set of monomers was then heated and solidified during space flight to produce the polymer, and the other set was heated and solidified on Earth.

After the experiment, the mechanical properties of the polymers grown under the two different conditions were measured. It was found that the microgravity environment can help produce polymers that are more uniform and gas permeable than those produced on Earth. A second set of experiments were conducted on mission STS-63 on the Shuttle *Discovery*.

The original experiments anticipated that commercial production of these polymers in space could begin as early as 1998. So are polymers for contact lenses produced in space now? The answer to that is no, not yet. Research has been hampered by access to facilities to conduct the experiments, scarce flight opportunities and only sporadic access to orbit. There are high hopes that access to the International Space Station will allow further experiments in space which will improve the materials and processes significantly. There are important steps still to be made in understanding how the polymers are formed.

Suppose that your research team believes that contact-lens polymer made in space has the necessary properties for semi-rigid lenses with high gas permeability, which improves the comfort to the wearer. Is it a sensible commercial proposition? In other words, can you reasonably expect investors to risk lending you money for several years against the chance of rich rewards? Note that the proposal is just to create the polymer in space. The lenses would subsequently be made from the polymer back on Earth.

Let's make some more assumptions in planning the project:

- The microgravity environment of the International Space Station is good enough for making contact-lens polymer.
- The manufacturing process in space should not require human intervention (astronaut time costs about £5 per second).
- A fully automatic manufacturing module can be developed for the International Space Station. It would carry 10 kg of lens material that could be processed over a period of six months.
- The polymer is sold for £1 million per kg (in this figure I'm including some of the profit after making and selling the lenses themselves, assuming that the actual lens manufacturing costs are relatively low and that people will pay a premium for better lenses that carry a 'Made in Space' logo).
- We can recover a batch of processed polymer, along with the manufacturing equipment for refurbishment and reuse.
- Staff are available as required and are not charged to the project when they are not needed.

Major cost items are summarized in Table 2.5. To work out the costs and income, we'll have to break the project down into phases of work.

Table 2.5 Costs associated with producing the polymer manufacturing equipment

Expense	*Cost*
Staff	£10k per person per quarter
Parabolic aircraft flights for testing	£5k each
Sounding rocket flights for testing	£25k each
Launch	£3000k each
Launch insurance	15% of the cost of replacing the manufacturing equipment

Phase 1 (quarters 1–4)

The first phase is to refine the processing of the new material in microgravity by means of experiments on-board an aircraft in parabolic flight. Parabolic flight is when an aircraft is flown to a high altitude at a steep angle, then the engines are idled and the plane traces a parabolic arc towards the ground. This gives 20–25 seconds of weightlessness on-board, as everything inside the aircraft accelerates towards the ground at the same rate. About 40 such periods can be obtained in each flight. Costs include 6 months with 3 people, 5 aircraft flights and £50 000 for materials and expenses.

Phase 2 (quarters 5–10)

The second phase will develop equipment for a ground test of the in-orbit manufacturing equipment, and qualify it for operation in space. Costs include 24 months with 12 people to develop and ground test, three months with 6 additional people to qualify, and £250 000 for materials and expenses.

Phase 3 (quarters 11–12)

The next phase will test the manufacturing equipment in microgravity, using aircraft in parabolic flight and sounding rockets. (Sounding rockets are rockets that launch the payload high into a parabolic arc, and return it to Earth on a parachute; several minutes of microgravity are experienced on such a flight.) Costs include 3 months with 6 people for 10 aircraft parabolic flight tests, 3 months with 6 people for 2 rocket flight tests, and £150 000 for materials and expenses.

Phase 4 (quarters 13–14)

The manufacturing equipment will be prepared for launch, insured, launched and commissioned in orbit. Costs include 3 months with 6 people for integration, and 3 months with 2 people for commissioning.

Phase 5 (quarters 15–16)

This is where value is finally added to the mission. The equipment will be run for 6 months by means of remote control from Earth. At the end of this phase the equipment and product will be recovered and returned to Earth (RTE). Costs include 6 months with 3 people for operation and launch and recovery charges of £3000k. The processed material is sold immediately after recovery.

Phase 6 (quarters 17–20)

The second cycle begins with refurbishing the equipment and repeating the production run. Costs include £150k for materials and 3 months with 6 people to refurbish; 3 months with 6 people for launch preparation and commissioning in orbit; 6 months with 3 people for operation and recovery; at the end of which the production batch is sold.

To replace the manufacturing equipment, if it was damaged or destroyed we would need 8 people for 12 months (half of the time to build the original equipment) and to spend another £250 000 for materials and expenses.

5.3.2 Project cash flow

With these assumptions, we can now plot the project's cash flow as the story unfolds. Like any development project, it starts off going negative (loss) as we invest in the development of the equipment and then (hopefully) goes positive as we start selling polymer. At this stage I'm ignoring any interest charges on negative balances. The plan is to borrow the money from investors on the promise of great returns after five years. I've also excluded the insurance charges for the time being.

The results of my spreadsheet work are shown in Figure 2.27. After just over four years, the project gets its first income and makes almost £2 million profit. The next flight generates additional profit so that, after five years, investors have over £5 million in the bank with all the investment repaid. Not a bad proposition?

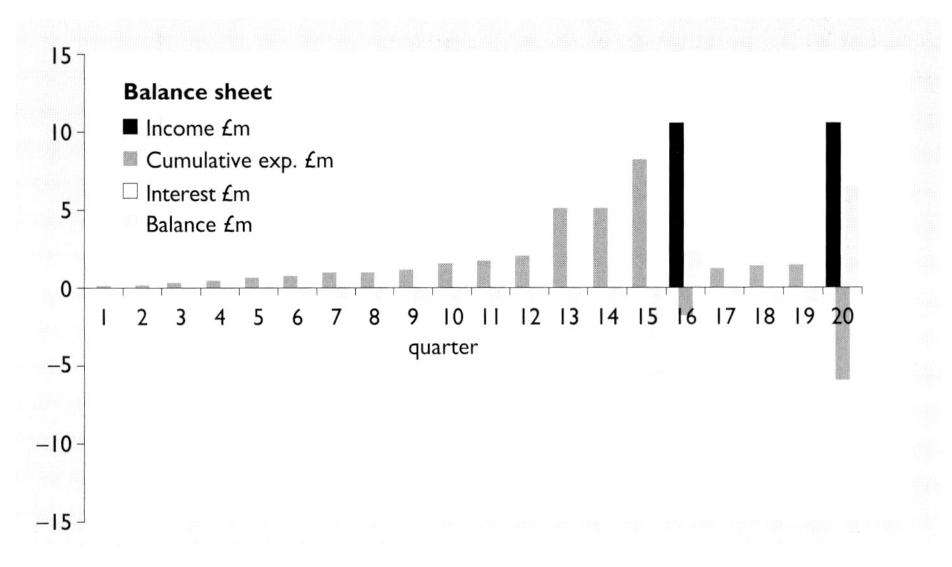

Figure 2.27 Cash flow of the contact lens polymer project, showing income (only when the polymer is sold); the cumulative expenditure (defined as total expenditure - total income); the interest on overdrafts (not shown here, but can be enabled on the spreadsheet for Activity 2.3); and the overall balance in the bank

5.3.3 Time value of money

The trouble with the cash flow in Figure 2.27 is the issue that we addressed earlier: the time value of money. Any commercial investor knows that money costs money, and wants to see a realistic return on investment. It's normal for venture capitalists to seek a return on investment (ROI) of 10%–30% per year for high-risk projects such as this one. After all, they could simply deposit their money in a bank where they might expect to receive interest of several per cent with near-zero risk. We can see the effect on the project by charging interest at the ROI on the outstanding debt.

Activity 2.3 Made in space

The spreadsheet used to generate the graph for Figure 2.27 is provided on the T173 CD-ROM in the section on spreadsheets (the relevant spreadsheet is entitled 'Polymer in space'). You will use the spreadsheet in the following exercises to investigate the viability of the proposed project.

Exercise 2.8

Using the spreadsheet, first set the value for insurance to 0%, then enter a value for interest charges of:

(a) 10%

(b) 20%

How does this affect the point at which a profit is generated?

For modest interest charges, it looks like that only after the second launch does the balance become positive. So who's worried? The investors might still be content, as they will have a steady return through interest charges; as operator, you are left with the liability until the balance after interest payment becomes positive. Only then can you actually pay back the investors.

When we ignored the need to include the ROI effect, the project was in profit after the first launch. With interest at 20%, though, the result is catastrophic with the balance rapidly plunging out of control.

The options available for making the project viable for such a high cost of capital would be a combination of some or all of the following:

- renegotiate the fixed charges for launches;
- reduce staff costs through productivity measures;
- raise the profit margins on the product by increasing the quantity processed per flight;
- raise the profit margins on the product by increasing the sale price of the product.

5.3.5 If disaster strikes

What happens if disaster strikes? It is not unknown for launch vehicles to fail spectacularly during lift-off. (Alternatively, the polymer-manufacturing unit could develop a fault during the launch which renders it useless, but ideally we will have ironed out any bugs during the sounding tests.)

If the first launch is a failure it will be necessary to claim on the insurance policy to rebuild the equipment and take a free flight from the launch contractor. Even at 0% interest, the one-year delay that might follow delays the time to profit. So perhaps we had better include some insurance premiums in our calculations.

The next exercise investigates this.

Exercise 2.9

Here are two tasks that examine the effects of insurance and a third one to illustrate where competition among the launchers could help.

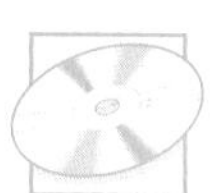

(a) *Zero interest with insurance.* The spreadsheet shows the balance sheet for the first five years up to the end of the second full cycle of manufacture. Set the interest rate to 0% and the insurance rate to 15%; this then adds 15% to the cumulative expenditure up to, but excluding, the first launch. What is the balance after the second batch of polymer is returned and sold?

(b) *10% interest with insurance.* Set the interest rate to 10% and the insurance rate to 15%.What is the balance now after the second batch of polymer is returned and sold?

(c) *The affordable launch.* Perhaps a rival consortium could handle the launch? Aim to break even after the second launch by reducing the launch cost. Set the interest rate to 10% and the insurance rate to 15% and find what launch price you could just afford if the balance is to be zero (within £0.1m) after the twentieth quarter.

(d) *Back into profit.* By how much does the launch cost have to be reduced in order for a £1.0m profit to be made after the second launch, using an interest rate of 10% and an insurance rate of 15%?

From the perspective of a venture capitalist, the result of including an insurance premium is disastrous, with the project still making a loss after the second launch, and only covering the interest payments in the meantime. You'd have to wait for the third launch to move into profit, and this sort of delay would probably cause a venture capitalist to drop the project.

This result leads to an interesting conclusion about insurance for a high-risk venture. Insurance underwriters often complain that new, entrepreneurial companies object strongly to insurance costs and are prepared to 'bet the company' on an uninsured launch. From our analysis, however, the venture is knocked out by a launch failure, insured or not. Insurance may not be as good a deal as it looks. Given that we know from the set-top box example that subsequent units are cheaper, perhaps the money would be better spent on developing a spare manufacturing plant.

5.3.6 Conclusions

What can we conclude from this simple analysis? The project looked good at first but as soon as we account for development costs, and, above all, the return on investment that a commercial investor would seek, it becomes a lot less appealing. The effects of any problem (launch failure or delay, marginal increase in costs) become much greater when the 'money costs money' effect is included.

This highlights one of the essential differences between a government-funded project and a private project. Government agencies, such as the military or the meteorological service, may trade off costs and benefits to establish whether a project is worth doing. But they are almost always able to work in terms of current costs and don't have to allow for the costs of debt. Indeed, they often prefer projects with flat spending profiles so that they can budget so much per year for the five years. The 0% ROI curves are often the ones they consider.

In comparison, the private investor looking for a profit has no difficulty finding other high-risk ventures offering a high return of 30% or more. Investors like projects with large expenditures pushed into the future rather than a flat spending profile.

This has great implications for the design and development procedures for a project involving space hardware and software. If a six-month delay at a late stage in development can lead to a two-year delay in returning profit to the investor, perhaps we shouldn't put so much effort into expensive and time-consuming space qualification of new components and systems.

Conventional thinking, which says that we should seek perfection in space projects, has no place in the grubby world of twenty-first century space venture capitalism!

6 Summary

In Part 1 of this block you looked at the process of designing. Some of the messages about design are quite general: that designs require constant evaluation, and that it is nearly impossible to set out a 'perfect' description (or model) of designing which will always generate a successful product.

In this part of the block, we have taken an example of one area of design, and looked at how the constraints of the field – high cost, high risk, high technology – influence both the design and the process of designing. Within this, you've seen that, because designers work in organizations, the way that an organization is structured can enable or stifle the production of a successful design.

In Part 1 we talked about the need to specify accurately the design problem. One of the issues that we discussed in the specification process was cost. When design is undertaken in a business context, it is geared towards producing a product that will make money. So some economics is needed in order to assess whether or not a new venture is worth the money. If the initial investment will never be recovered, then there is no point in proceeding beyond the concept stage. The example of making set-top boxes showed a simplified cash-flow analysis that could be used to assess the future profits from a product. The set-top boxes have the advantage that they can have one or more prototype stages during the development: the first ones off the production line can be used to make refinements also. The luxury of a prototype is almost unheard of for space applications, so recouping the development costs and making a profit may ride on the success of a single item. In this case, hard financial realities can play a key role in influencing both the design itself and the timescale over which the design can be progressed.

A common thread running through both parts of this block has been the idea of modelling. We saw in Part 1 how designers use models to develop their ideas, or to show other people what their ideas are. A model could be a simple concept sketch, such as the Aston Martin sketch we saw in Figure 1.6(a), or an architectural model, as in Figure 1.3, or a finite-element computer model, as in Figure 1.7. The essential point is that a model is a simplified representation of reality.

There are many kinds of model. In Part 2 we made some simple assumptions and estimates relating to a space-based business proposal, and looked at the effect that changing those assumptions had on the viability of the project. This too is a kind of model. Engineers routinely use this sort of model in their work. Engineers' models might relate to the viability of a project, or to its technical aspects – such as the effects of using one material rather than another, or one manufacturing process instead of another. The ramifications of these choices could be explored, possibly using computer tools such as spreadsheets, or perhaps using approximate, back-of-envelope calculations. In fact, modelling is almost inescapable once we start to think speculatively about what might be done, or should be done, or could be done in the physical world. This is very much the province of the engineer.

7 Learning outcomes

After studying this part of Block 2 you should:

2.1 Understand the possibilities that are offered by space for exploration and commercialization.

2.2 Understand that the cost of launching a satellite or spacecraft from the Earth can be affected by the design of the launcher used.

2.3 Understand that organizational structures within companies can affect the way in which a design project is managed.

2.4 Recognize the importance of considering the appropriate system, rather than just the individual component, when designing or making decisions.

2.5 Recognize that the term 'system' refers to a particular categorization of the components or factors in a situation, and that different individuals can have different conceptions of the system that is relevant.

2.6 Given appropriate information, use a multiple-cause diagram to represent the causal interrelationships between factors in a particular situation.

2.7 Given appropriate information, use a systems map to portray the structure of the system perceived in a particular situation by individuals or groups with different perspectives.

2.8 Understand that the design process can be influenced by factors such as enforced cost reduction and the consequences of product failure.

2.9 Recognize that many engineering objects have moved from being driven by new technology to being driven by the needs of the 'market'.

2.10 Interpret information provided in graphical format, and plot data points onto a graph with linear axes.

2.11 Understand how spreadsheets may be used for simple modelling of a business venture.

Answers to exercises

Exercise 2.1

Light travels at 3×10^8 ms^{-1}. An elapsed time of 10.5 hours corresponds to:

$$10.5 \times 60 \times 60 \text{ s} = 37\,800 \text{ s}$$

So Pioneer 10 was at a distance of:

$$3 \times 10^8 \text{ m s}^{-1} \times 37\,800 \text{ s} = 1.134 \times 10^{13} \text{ m}$$

Or just over 11 trillion metres!

Exercise 2.2

(a) Kennedy set the objective of landing a person on the Moon.

(b) The constraints are the difficulty of the venture, and lack of knowledge of the benefits of the project.

(c) The criteria for success are landing a person on the Moon, and returning that person safely to Earth. It could be argued that this is the same as the objective of the project. However, a project can be successful and yet not achieve its final goal. So landing on the Moon might be seen as a success even if the astronaut never returned!

Exercise 2.3

The first 55 have been produced with a total outlay of £500 000. So the cost of each is £500 000 divided by 55, which works out at about £9000 per unit. (Note that the competition is selling units at about £200 each.)

Exercise 2.4

The company makes a profit in the eighth quarter.

Exercise 2.5

(a) Raising the staff cost to £14k delays the profit until the 11th quarter.

(b) I found a unit price of £210 (5% rise) was just enough to meet the 40% increase in staff costs. (There is only a very small loss at this price.)

Exercise 2.6

It is delayed by one quarter.

Exercise 2.7

£300, or £0.3k.

Exercise 2.8

(a) For modest interest charges at 10%, the project's revised cash flow shows that only after the second launch does the balance become positive at a little over half a million pounds.

(b) When I tried running the spreadsheet with interest at 20% the result was catastrophic with the balance rapidly plunging out of control. The proposition is starting to look less attractive.

Exercise 2.9

(a) The balance after the second launch is just over £5 million (£5.37m): not too much reduced.

(b) The balance is now negative after the second launch, by about £0.7m.

(c) I found that at £2.9m per launch the loss was a mere £100k. That's a reduction in launch cost of just over 3%.

(d) Reducing the launch cost by 10%, to £2.7m per launch, will give a profit of around £1m.

Answers to self-assessment questions

SAQ 2.1

(a) Space exploration in the next few years will see robotic missions to Mars (and perhaps a manned mission a little further ahead?).

(b) Looking further into the future there will be missions to the moons of Jupiter and Saturn.

(c) Space commercialization in the short term may primarily revolve around gimmicks such as space tourism and orbital creation.

(d) Further in the future we may see manufacturing plants in space and mining of the Moon or asteroids for mineral resources.

SAQ 2.2

(a) Launchers that have larger payload capacities offer cheaper costs per kg of payload. Figure 2.7 shows that it is the heavier launchers which can carry the larger payloads. So in designing a launcher for cheap flights, it looks like bigger is better. The people responsible for the launch vehicles should presumably try to ensure that the launcher is filled to capacity. Note that the cost is given *per kg*, so a heavier satellite will still cost more to launch than a light one.

(b) If the data is added to Figure 2.6(a), the result is Figure 2.28.

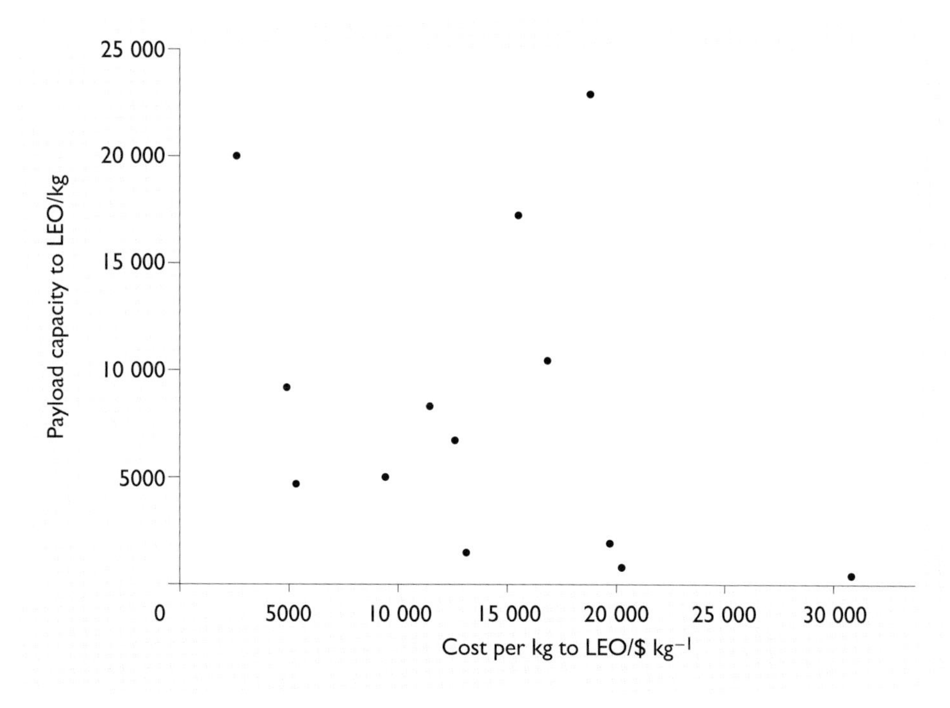

Figure 2.28

So the Shuttle and the Titan are more expensive than other launchers of comparable capacity. Including these two makes it difficult to identify a trend as easily as in Figure 2.6.

(c) As these launchers are used for military missions, where the cost may be of little object, they can afford to be less cost-efficient. The shuttle is also expensive because of its reusable nature.

SAQ 2.3

See Figure 2.29.

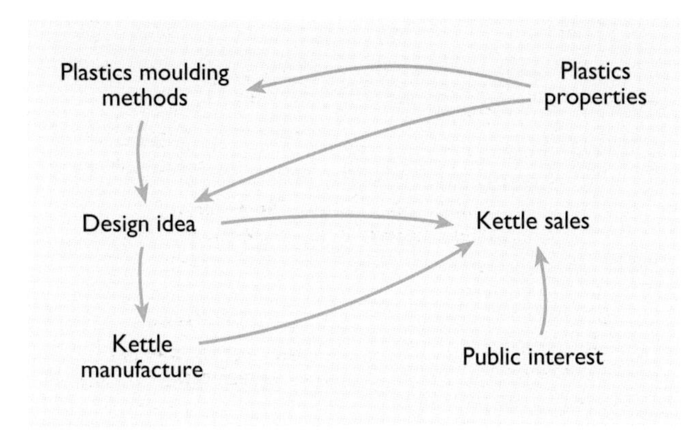

Figure 2.29 A multiple-cause diagram of the additional factors affecting jug kettle sales

SAQ 2.4

See Figure 2.30.

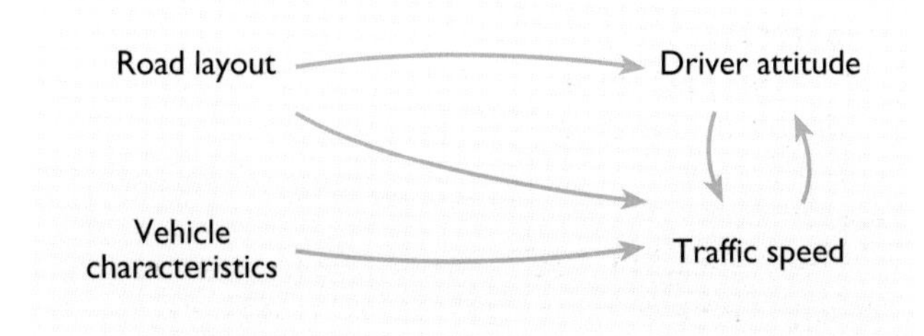

Figure 2.30 A possible multiple-cause diagram to show the relationship of the factors affecting traffic speed on a stretch of road with traffic calming measures in place

SAQ 2.5

See Figure 2.31.

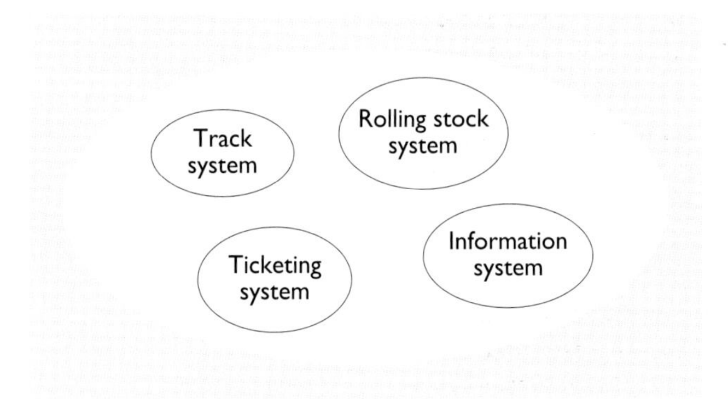

Figure 2.31 A systems map of the railway transport system

SAQ 2.6

See Figure 2.32. Note that there are a lot of similarities between the two diagrams, as you might expect, but there are important differences. So, in both cases, the person considering the system would need to take into account the supply of power on the station, the ways in which the station's operation is

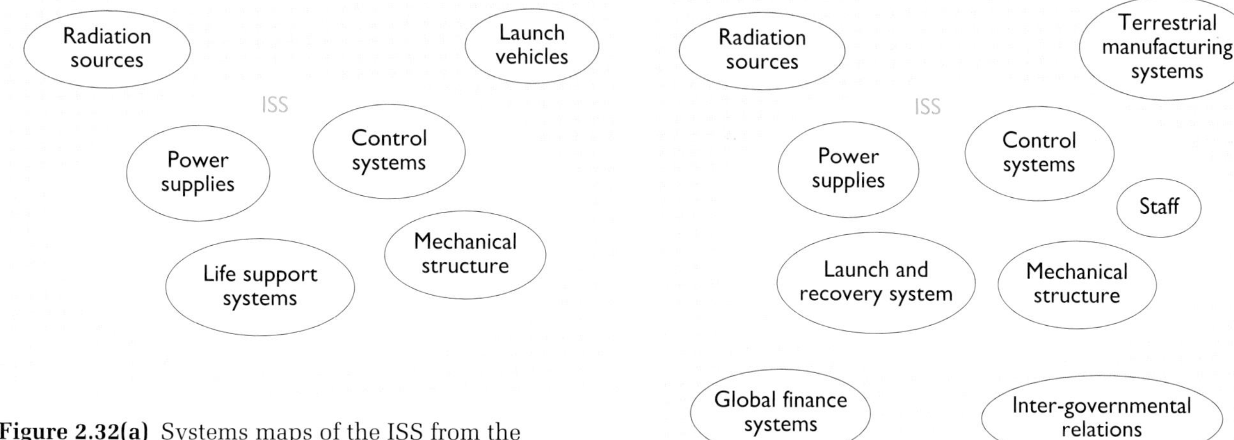

Figure 2.32(a) Systems maps of the ISS from the perspective of a company wishing to design one of the service components

Figure 2.32(b) Systems maps of the ISS from the perspective of a company considering investing in an experimental manufacturing process on the ISS

controlled, and its mechanical structure. The radiation sources in the environment (and possibly sources of microdebris that could damage the station) are also relevant to both. For a component designer, the launch vehicle system is also part of the environment, but for the manufacturing company, the launch/recovery system could be regarded as a vital part of the ISS, since without it, manufacturing in space would be pretty pointless. For the component manufacturer, the life support system on the ISS would probably be regarded as a major part of the system, whereas for the manufacturer, it is probably the skills and time available from the station staff that would be more directly important. For the manufacturing company, financial systems, competing terrestrial manufacturing opportunities and the relationships between the governments involved in the ISS would probably be important considerations that form part of the overall environment.

SAQ 2.7

(a) This is clearly best suited to a matrix organization. The project is pulling expertise from several sources within the company, so a functional organization is not appropriate. However, it will benefit the company if staff expertise is shared between duties on different projects. Hence the matrix approach is best. Whether a strong or weak matrix should be used is something I leave for you to decide!

(b) In this case a functional-organization type is best. The manufacturing company need have no input from anything other than its manufacturing engineers.

SAQ 2.8

(a) Exploration using astronauts is costly, because of the extra equipment associated with life-support on a long mission, and because failure cannot be tolerated, as it leads to loss of life. So this falls into our category of a 'vicious circle' of space design.

In a series of missions, failure of a single spacecraft should not wreck the entire programme. Also, each individual mission can be relatively inexpensive. So this is part of the 'virtuous circle'.

(b) Space missions have become less driven by 'technology push' and are more influenced by 'market pull'. So the cost of a project needs to show a return in terms of scientific knowledge or data, and not just be spent on a showcase of cutting edge technology where the benefits of new developments may not reach the public for several years.

References

Clarke, A. C. (1945) 'Extraterrestrial Relays', *Wireless World*, October.

Baker, D. (2000) *Scientific American: Inventions from Outer Space*, Random House, New York.

Acknowledgements

Grateful acknowledgement is made to the following sources for permission to reproduce material in this book.

Part 1

Figures

Figure 1.1: From left to right: © Science Photo Library, © Francoise Sauze/ Science Photo Library, © Jerome Yeats/Science Photo Library. © Jerome Yeats/Science Photo Library; *Figure 1.3:* © Edifice/Norman; *Figure 1.4:* © Christie's Images; *Figure 1.5:* © Hulton Getty; *Figure 1.6:* Courtesy of Aston Martin Lagonda, Newport Pagnell, Buckinghamshire; *Figures 1.9 and 1.13:* Courtesy of Pifco Holdings plc, Manchester; *Figure 1.24:* Reproduced by permission of the A.O.C. and Commandant Royal Air Force College, Cranwell; *Figure 1.28:* Gossamer Albatross © NASA, 1980; *Figures 1.29 and 1.30:* © Courtesy of Rolls-Royce plc; *Figures 1.37–1.43:* © R. I. Davidson; *Figures 1.44, 1.49, 1.52:* © Mike Hessey, mike@whooper.demon.co.uk; *Figure 1.48:* BSA folded/unfolded, Dahon folded/unfolded, Dursley Pederson folded/unfolded, Airframe folded/unfolded © Mike Hessey mike@whooper.demon.co.uk; Trusty Spacemaster folded/unfolded © Tony Hadland; Faun, from a Faun catalogue, 1895 and Grout patented in 1880 reproduced from *It's in the bag! A history in outline of portable cycles in the UK*, by Tony Hadland and John Pinkerton. Published by Dorothy Pinkerton, Birmingham (1996); *Figure 1.61:* View of Avon Gorge with the approved design for the Clifton Suspension Bridge, 1831 by Samuel R.W.S. Jackson (1794–1868). Bristol City Museum and Art Gallery, UK/Bridgeman Art Library; *Figure 1.62:* Reproduced with the permission of the Librarian, the University of Bristol; *Figure 1.63:* Courtesy of Places and Spaces, London SW4.

Thanks also to Tim Gamble, Mark Endean and Karen Kear of the Open University for the loan of their bikes for Figures 1.46, 1.50 and 1.51 respectively.

Text

Page 69: 'Kew for a ride', *The Standard,* February 3, 1982, © 1982 Evening Standard; *Page 75:* Brown, T. (1997), 'Nurturing a culture of innovation', *Financial Times,* 17 November, 1997, © 1997 Ideo Europe.

Part 2

Figures

Figure 2.1: © NASA/Still Pictures; *Figures 2.2, 2.5 and 2.13:* © NASA.

Every effort has been made to trace copyright owners, but if any have been inadvertently overlooked, the publishers will be pleased to make the necessary arrangements at the earliest opportunity.

T173 Course Team

The following course-team members were responsible for this block.

Academic Staff

Dr Michael Fitzpatrick (Course Team Chair)
Adrian Demaid
Professor Chris Earl
Steve Garner
Dr Jeff Johnson
Dr Dick Morris
Professor Nicholas Braithwaite
Dr Bill Kennedy
James Moffatt
Dr George Weidmann
Jim Flood
Dr Suresh Nesaratnam
Professor Bill Plumbridge
Professor Robin Roy

Consultants

Rodney Buckland
Dr Sarah Hainsworth

Production Staff

Sylvan Bentley (Picture Research)
Philippa Broadbent (Materials Procurement)
Daphne Cross (Materials Procurement)
Tony Duggan (Project Control)
Elsie Frost (Course Team Secretary)
Andy Harding (Course Manager)
Richard Hoyle (Designer)
Allan Jones (Editor)
Lara Mynors (Editor)
Howie Twiner (Graphic Artist)